Reify = thing-iF-ICA

abstract = concrete

Apple & Gravity — Gravity a property of Apples,
since they Always FALL.

Gravity is a concept which describes Relationship
Bet 2 unequal masses.

SOCIOBIOLOGY EXAMINED

~~problem~~ note with SB - thing is "inside" the body &
∴ intervention is possible.

But how can me "cut out" a thing / concept.

———

Wilson is A positivist - ie - he wants to reduce all concepts
statements of to statements of FACT.

~~~~~~~

Falsifiability —

Similarly — there is no such "thing" as
intelligence - if you mean a thing located in
~~the~~ brain which produces certain Behaviour
Gould's Point - this Reification is logically similar
to saying gravity is in the apple.

Contributors

Jerome H. Barkow
DALHOUSIE UNIVERSITY

S. A. Barnett
AUSTRALIAN NATIONAL UNIVERSITY

Derek Freeman
AUSTRALIAN NATIONAL UNIVERSITY

Stephen Jay Gould
HARVARD UNIVERSITY

Marvin Harris
COLUMBIA UNIVERSITY

James C. King
NEW YORK UNIVERSITY SCHOOL OF MEDICINE

David Layzer
HARVARD UNIVERSITY

N. J. Mackintosh
UNIVERSITY OF SUSSEX

Mary Midgley
UNIVERSITY OF NEWCASTLE UPON TYNE

Ashley Montagu
PRINCETON, NEW JERSEY

Karl Peter
SIMON FRASER UNIVERSITY

Nicholas Petryszak
UNIVERSITY OF ALBERTA

Steven Rose
OPEN UNIVERSITY

Thomas Sheehan
LOYOLA UNIVERSITY OF CHICAGO

Michael A. Simon
STATE UNIVERSITY OF NEW YORK AT STONY BROOK

S. L. Washburn
UNIVERSITY OF CALIFORNIA AT BERKELEY

SOCIOBIOLOGY EXAMINED

Edited by
Ashley Montagu

OXFORD UNIVERSITY PRESS
Oxford New York Toronto Melbourne
1980

Oxford University Press
Oxford London Glasgow
New York Toronto Melbourne Wellington
Nairobi Dar es Salaam Cape Town
Kuala Lumpur Singapore Jakarta Hong Kong Tokyo
Delhi Bombay Calcutta Madras Karachi

First published by Oxford University Press, New York, 1980
First issued as an Oxford University Press paperback, 1980

Library of Congress Cataloging in Publication Data
Main entry under title:

Sociobiology examined.

    Includes bibliographies.
        1. Sociobiology—Addresses, essays, lectures.
I. Montagu, Ashley, 1905–
GN365.9.S63   591.5   79-21169
ISBN 0-19-502711-6
ISBN 0-19-502712-4 pbk.

Printed in the United States of America

Dedicated
to
LEWIS MUMFORD

Contents

SOCIOBIOLOGY EXAMINED

INTRODUCTION

When, in 1975, Edward O. Wilson's book *Sociobiology: The New Synthesis* appeared, it was greeted with the kind of attention that few books of its kind have received in our time. Except for a relatively few dissenters it was widely and favorably reviewed in the scientific, literary, and lay press. The book created something of a sensation principally because of its claim to have laid the foundations for a new science, one that would for the first time provide a firm biological basis for the understanding of the refractory human social behavior with which social scientists have ineffectually attempted to grapple for too long a time. Having written a quite admirable 546 pages of beautifully illustrated text setting out the principles of the new discipline of sociobiology—which Wilson defines as the systematic study of the biological basis of all forms of social behavior, including sexual and parental behavior, in all kinds of organisms, including man[1]—Wilson devotes his concluding chapter to "Man: From Sociobiology to Sociology." It is this twenty-seventh chapter which has given rise to much criticism and a continuing debate which promises to be as lively as any in recent decades. I am familiar with at least a hundred articles on the subject and more than half a dozen full-sized books and anthologies on sociobiology. There have been debates on the subject on radio and TV, and conferences specially devoted to sociobiology, not to mention the innumerable lectures delivered on it.

Unfortunately some of the discussion has been marred by a bias and an occasional intemperateness which has only served to lose sympathy for the guilty ones. Wilson has withstood these assaults upon his integrity with civility and the appropriate sense of humor. Truth will not be advanced by heat so much as it will by light. Nor will it be advanced by the kind of loose thinking and blinkered ignorance that has characterized some of the principal participants in the debate. It is a matter of general experience that where nature-nurture issues are concerned, emotion only too often enters in. It is strange to see the "mature" "dispassionate scientist" behaving in this way only because the "mature" "dispassionate scientist" is a myth. Scientists are human beings with their full complement of emotions and prejudices, and their emotions and prejudices often influence the way they do their science. This was first clearly brought out in a study by Professor Nicholas Pastore, *The Nature-Nurture Controversy* (New York: King's Crown Press), published in 1949. In this study Professor Pastore showed that the scientist's political beliefs were highly correlated with what he believed about the roles played by nature and nurture in the development of the person. Those holding conservative political views strongly tended to believe in the power of genes over environment. Those subscribing to more liberal views tended to believe in the power of environment over genes. One distinguished scientist (who happened to be a teacher of mine) when young was a socialist and environmentalist but toward middle age he became politically conservative and a firm believer in the supremacy of genes!

Today we see a similar division between those who are politically left and those who are politically right of center. I mention these matters because I think they are important, and because in spite of all attempts to disavow the role that political views and private emotions have played in the presentation of sociobiological ideas and their criticism it is clear that such biases have frequently been involved. It is well to remember this in reading any discussion of sociobiology—for or against.

As will become evident from a reading of the individual contributions to this volume, Wilson's biases and prejudices are especially evident in his application of sociobiology to humankind. And as in many another instance, Wilson seems to be quite unaware of the manner in which he has been influenced by his personal views in presenting what he considers to be his scientifically based arguments. This begins with the very first sentence with which he opens his concluding chapter. "Let us now consider," he writes, "man in the free spirit of natural history, as though we were zoologists from another planet completing a catalog of social species on Earth."[1] But that is impossible. It cannot be done. One cannot divest oneself of one's conditionings and apperceptions. There are no primeval Adams, and in spite of the flying saucer enthusiasts, no visitors—not even zoologists—from other planets. As approximations to rational beings it is not for us to indulge in pseudological rationalizations or metaphrastic translations of extra-terrestrial zoologists into mouthpieces for our very terrestrial views, but rather to acknowledge our biases and prejudices and do everything in our power to guard against them.

Wilson is a distinguished scientist whose specialty is entomology. He has written a book *The Insect Societies,* which is already a classic. He has produced a great deal of original work based on his researches which are of fundamental importance. And he has obviously read widely and deeply in the literature of natural history. He writes extremely well. He writes interestingly and holds the reader's attention. But his overweening fault is biologism, his conviction that since humans are animals who have evolved in much the same ways as other animals they must be explicable in much the same way. Hence, in the very next sentence following that which I have already quoted, Wilson writes, "In this macroscopic view the humanities and social sciences shrink to specialized branches of biology; history, biography, and fiction are the research protocols of human ethology, and anthropology and sociology together constitute the sociobiology of a single primate species." In an article, published the following year, entitled "Sociobiology: A New Approach to Understanding the Basis of Human Nature" (*New Scientist,* May 13, 1976, pp. 342–44), Wilson emphasizes the point. "The role of sociobiology," he writes, "with reference to human beings, then, is to place the social sciences within a biological framework, a framework constructed from a synthesis of evolutionary studies, genetics, population biology, ecology, animal behaviour, psychology and anthropology."[2]

This, in my view, is to turn things topsy-turvy. Contrary to

Wilson, I would place all the sciences relating to humans within an anthropological framework—where, surely, they naturally belong—and make sociobiology an intrinsic part of that framework. A sound sociobiology has a genuinely useful role to play in the explication of the social behavior of animals. In studying and comparing the societies of different kinds of animals and of humans it is probable that some light may be thrown on the evolutionary origins of some forms of social behavior. The aim, says Wilson, is to construct and test theories about the underlying hereditary basis of social behavior. That, surely, is an endeavor which should receive nothing but encouragement. The danger, however, in such an endeavor arises from the attempts to prove rather than to disprove the assumptions generally made that behavior, social behavior, has a certain underlying hereditary basis.

What many sociobiologists fail to understand is that every observation is an experiment, a fact pointed out long ago by F. C. S. Schiller,[3] and that the manner in which we make our observations often determines what we will see. A besetting sin of sociobiologists is that they are prone to start out with the assumption that they are going to find hereditary bases for the social behavior they observe, and invariably succeed, by analogy or extrapolation or misinterpretation, in confirming their anticipated findings. I think there can be not the slightest doubt that a good deal of human social behavior has a genetic basis, but that is a very different thing from claiming that such behavior is genetically determined. And yet sociobiologists do tend to write as if genes play a major role in determining human social behavior. As illustrative of this tendency Wilson, for example, writes that one of the "sophisticated" ways in which genes replicate themselves is through altruistic behavior. Within any group its reproductive efficiency or chances of survival may be increased by the altruistic act of a group member. If the group members are genetically related then their genes are likely to be transmitted to subsequent generations by the altruistic act of one of the members. "Natural selection," writes Wilson, "will therefore select favourably for such altruistic acts, and thus for the genes that determine them."[4]

But do genes "determine" altruistic acts? They may do so in insects on which Wilson is an authority, but as an anthropologist

I consider it more than doubtful that they do so in humans. Surely, common experience tells us that some individuals are quite incapable of an altruistic act, and that variability in this is as great as it is in any other human behavior? Wilson acknowledges that human social evolution has been more cultural than genetic. Nevertheless he considers that the underlying emotion of altruism, "powerfully expressed in virtually all human societies," is the consequence of genetic endowment.

But as Harlow and his co-workers have shown, monkeys who have been isolated or inadequately socialized are, in later life, altogether wanting in anything resembling a capacity for altruistic behavior.[5] The same is true of humans who have suffered similar deprivations in infancy and childhood. That altruistic behavior has a genetic basis I have not the least doubt. I have repeatedly set out the evidence for this, and it has recently been confirmed in babies and infants whose altruistic behavior has long been known to some if not to others. What is, however, clear is that environmental factors play a decisive role in determining whether such behaviors will be developed or not.

Making some very sensible comments on the unlikelihood of aggression being an "instinct" in birds and mammals, Wilson suggests that the key to aggression is environment. He then proceeds to the discussion of what he considers to be the crucial issue with which sociobiology has to grapple, namely, the relative contribution to human behavior of genetic endowment and environmental experience. "It seems to me," Wilson writes, "that we are dealing with a genetically inherited array of possibilities, some of which are shared with other animals, some not, which are then expressed to different degrees depending on environment."[6] This statement is, of course, quite unexceptionable. What is exceptionable is Wilson's marked tendency to give more weight than they deserve to genetic influences that may be involved in various behaviors. For example, Wilson suggests that "the tendency under certain conditions to indulge in warfare against competing groups may well be in our genes, having been advantageous to our Neolithic ancestors."[7] Since the gatherer-hunter peoples, like the Australian aborigines, the Bushmen of Africa, the pygmies of the Congo, the Andaman Islanders, the Eskimo, and others, never achieved a Neolithic stage of development, and since they are not given to warfare, one wonders whether Wilson would exempt them from the flawed genes which he conjectures may be responsible for the tendency, "under certain conditions," to make war, or whether he considers that those genes began to flourish in the Neolithic because they proved advantageous to our ancestors? The truth is that gatherer-hunter peoples do not engage in warfare for the simple reason that they have no reason to do so. And that, surely, presents a very evident and significant point, namely, that if evidences of warfare first appear in the Neolithic, then it is to "certain conditions" that we should look for the efficient causes of war.

But do evidences of warfare first appear in the Neolithic? I am not aware that they do. There is no unambiguous evidence of warfare during that stage of cultural development. Implements exist that might have been used in the chase or in war. We have a lot of evidence for the former, but none at all for the latter. And even if such evidence were available, what kind of evidence would that constitute for a genetic substrate for such behavior? The answer is none. It is the merest conjecture, wholly unsupported by any fact, that warfare is in any way fuelled by genes.

The very best and the latest examination of the roots of war is Richard Barnet's book of that title, in which virtually every aspect of the subject is considered. Barnet concludes:

> Human nature makes war possible, but it is not its cause. Despite efforts of the apologists of American policy to justify permanent war by invoking pop anthropology, the view that militarism is a biologically determined aspect of the human condition cannot stand serious scrutiny. . . . The dismal view of man as a natural warrior offers a certain bizarre comfort because it absolves individuals of responsibility of identifying, much less removing, the specific political, organizational, and economic causes of militarism and war. If human beings, males particularly, have biological urges to slaughter their own species at regular intervals, there is nothing to be done. Remorse is as useless as reform.[8]

Such has been the conclusion of all serious students of the subject.

"Sociobiology" writes Wilson, "can help us understand the basics of human behaviour and the fundamental rules that govern our potential."[9] Apparently Wilson assumes these rules to be basically genetically influenced, for he goes on to say, "We will need to know how, genetically, certain types of behaviour are linked to others." Is there any evidence that certain types of behavior are genetically linked to others; I do not know of any. Judging from experience it seems to me that every human being must learn each of his behaviors for himself; that the individual's behavior is largely custom-made, modelled after the patterns prevailing in his particular group. What sociobiologists do not fully understand is that as a consequence of the unique history of human evolution humankind has moved into a completely new zone of adaptation, namely, culture, the human-made part of the environment; that it is through the learned part of the environment that humans respond to the challenges of their environments, and not through the determinative or decisive action of genes. Yet, Wilson, in his latest book, *On Human Nature* (1978), can write, "The question of interest is no longer whether human social behavior is genetically determined, it is to what extent. The accumulated evidence for a large hereditary component is more detailed and compelling than most persons, including even geneticists, realize. I will go further: it already is decisive."[10]

And what is the evidence for this genetic determinance? Wilson offers as examples the size of adult groupings on the order of ten to one hundred adults, unlike birds and marmosets with just two, or thousands as in insects and fishes. The larger size of males, giving the greater advantage to larger males in the competition for females. The long period of social training of the young with their mothers. Play.

When we examine the actual evidence provided by existing human societies we find no support whatever for "a large hereditary component" in such human social behaviors. Professor Marvin Harris, in his contribution has dealt so thoroughly with these wholly unsupportable claims of Wilson I need say no more here.

What Wilson fails to understand is that the universality of a trait in a species like *Homo sapiens* does not constitute evidence of genetic determination. And certainly he provides no genetic evidence for the genetic claims that he makes. Common environmental influences in humans are likely to produce common forms of social behavior. Given human capacities, it is hardly necessary to introduce the *deus ex machina* of genetic determination in order to explain such behavior. As Professor Michael Simon says in his contribution, "It is very likely true that whatever is universal in humans is biologically significant, in the sense of contributing to the perpetuation of the species, but it does not follow that any of these qualities must be determined genetically."

Professor Simon has many other wise things to say on this subject.

Furthermore, the simplistic manner in which Wilson speaks of the "genetic" or "hereditary component" in human social behavior betrays the fact that his genetics is rather dated. As Professor James King shows in his contribution, Wilson's genetics, in which selection acts on single genes, and genes are viewed as discrete units, which vary only in quantum mutational jumps, is somewhat outmoded, to put it generously. The researches of recent years indicate that natural selection acts not so much on single genes as on gene combinations, as King says, "in individual chromosomes such that when these chromosomes are randomly combined to produce new individuals, the new combinations will have a high probability of producing individuals of the modal phenotype." Indeed, evolution by the action of selection on single genes is probably a rare occurrence.

The revolution in our thinking that is under way concerning these and other genetic matters is admirably discussed by Professor King; the consequences for sociobiological single gene genetics are devastating.

However much they may be willing to acknowledge the influence of learning or culture, sociobiologists nevertheless continue to write as if it were genes that determine the patterns of human social behavior—a notion that founders on the fact of the unique educability of humankind. Not a *tabula rasa* by any means, but full of potentialities which do indeed have a firm genetic basis, but which are completely malleable under the shaping influences of culture. Wilson pays lip-service to the interaction between genes and environment, and then forgets about interactionism in the pursuit of his argument. What is worse, he fails to understand that, as Lehrman pointed out many years ago, the interaction out of

ORganism + environment
(not gene)

which the organism develops is not between genes and environment, but between organism and environment, an organism that is different at each stage of development.[11] In short, it is the action of the environment upon the organism that influences and makes possible the functional expression of the genes. The genes that influence behavior are fortunately not the equivalent of predestination. If there is one trait which more than any other distinguishes *Homo sapiens* from all other living creatures it is educability. Educability is the species trait of humankind. Humans are polymorphously educable, which is to say that they are capable of learning anything and everything it is possible to learn. Under the pressures of the environmental challenges in which humankind has developed it is plasticity, flexibility, and malleability that have been at a premium. In the environments in which humans have developed, the pressures of natural selection have been not upon the development of any special trait or capacity—always, of course, excluding the special capacity for speech—as upon generalized problem-solving ability or intelligence, the capacity to develop *responses* to any and every challenge of the environment. The operative word here is *responses,* not reactions—the ability to make the appropriately successful response to the challenge of the particular situation. That is the definition of intelligence.

Just as the size of the motor area of the brain which controls various muscles does not relate to the size of the muscle but to the *skill* in using it, so the size of the human brain, and especially of the cortex, relates not so much to the special functions with which it is concerned but to the skill with which it is capable of functioning as an organ of inhibition, learning, choice, response, information-processing, and analyzing. The skill with which humans can do all these things far exceeds that of which any other animal is capable. To say that this kind of a brain is equivalent to a *tabula rasa* would be quite wrong, for its potentialities for the skills which it is capable of developing undoubtedly have some sort of a basis in interacting genes. It is the uniqueness of the human brain that must be emphasized, the fact that it is the organ that permits voluntary rather than biologically predetermined behavior. It is the organ of deliberation and choice. As José Delgado, the distinguished neurophysiologist has said, to behave is to choose one pattern among many. It is also the ability to maximize the improbable, to transform accidents into opportunities, to innovate, to be opportunistic. This, again, is a species characteristic of our kind, almost wholly the consequence of what the species can do with its genetically based educability under the influence of the environment. The advantages of this computer-like versatile problem-solving brain, as Professor King says, is that rigid ready-made answers do not have to be built into it. "On the basis of general rules and stored information gleaned from experience it can generate novel ways of solving problems. Once this type of plasticity reaches a certain level, it becomes more adaptive for selection to increase the plasticity and the versatility and to reduce the number of rigid, ready-made answers built into the program. It seems much more likely that selection has worked toward greater flexibility in human behavior rather than toward amassing a great aggregation of rigid imperative minutiae."

It is that kind of brain that makes human beings what they are, and that kind of genetic system which makes it possible for them to become the kind of innovative creatures they so conspicuously are, and it is these facts that constitute the final answer to the claims of the sociobiologists.

And now to another matter. Ideas, as is generally known, have consequences. The idea of "race," for example, led directly to the holocaust. Early in this century it was predicted that it would. [12, 13] Inherent in the idea of sociobiology as presented by Wilson, and by some of his more extreme followers, such as Richard Dawkins (whose preposterous book *The Selfish Gene* is anatomized by Dr. Mary Midgley in Chapter 5) is the notion of the genetic determinism of behavior of individual differences, social differences, the stratification of classes, sexual status, and racism. I doubt whether Wilson, Dawkins, or any other sociobiologist consciously intended any such thing, but whatever their personal virtues may be, as Dr. Midgley says, "Sociobiology as a movement is a real menace, because it provides simple-minded people who like the jargon of science with an exceptionally slick set of catchwords and formulae for universal explanation. Like any flag-waving movement, as is gathers strength it is bound to collect a mass of supporters who will catch their leader's confidence without his scruples and without understanding his limitations. . . . the academic world is full of people who ask nothing better than to settle into such an army" (Chapter 2, p. 26). Indeed, such an army is already with us, recruited from virtually every academic discipline as well as right-wing political groups. On an intellectual level France seems always to be fertile soil for the seeds of any ideas that can be cultivated for reactionary purposes. Sociobiology has proved a godsend to these neo-fascist New Rightists. Alain de Benoist, their principal spokesman, and cultural editor of *Le Figaro* magazine, declared "The enemy is not 'the left' or 'Communism' or 'subversion' but the egalitarian ideology whose formulas . . . have flourished for 2,000 years." With the rejection of egalitarianism go the democratic ideals that flow from it. The New Rightists hold that individuals and races are insurmountably separated from one another by hereditary inequality; this view is buttressed by citing the writings of Wilson, William Shockley, and Arthur Jensen. But Professor Thomas Sheehan gives so thorough an account of these developments in France in Chapter 16 I shall not detain the reader further here.

In England the organ of the fascistic National Front, *Spearhead*, has featured several articles on sociobiology. One of them, written by Richard Verrall, entitled "Sociobiology: The Instincts in our Genes" (March 1979, pp. 10–11), informs the reader that sociobiology has shown that "there exist instincts which are genetic in origin and which determine our behaviour and social customs"; that "sociobiology is thus transforming our view of man and society," but that "Of a far greater significance is the basic instinct common to all species to identify only with one's like group; to in-breed and to shun out-breeding. In human society this instinct is *racial*, and it—above all else—operates to ensure genetic survival," and much else to the same effect.

I am not aware that any sociobiologist has taken the trouble to repudiate such misuse of his ideas.

Finally, let me urge upon the reader the specially careful reading of Dr. Mary Midgley's fine contribution to the methodological issues involved in the sociobiology controversy, "Rival Fatalisms: The Hollowness of the Sociobiology Debate." It should give many of us cause furiously to think.

NOTES

1. Edward O. Wilson, *Sociobiology: The New Synthesis*. Cambridge: Harvard University Press, 1975, p. 547.
2. Edward O. Wilson, "Sociobiology: A New Approach to Understanding the Basis of Human Nature." *New Scientist*, May 13, 1976, p. 342.
3. F. C. S. Schiller, "Scientific Discovery and Logical Proof." *In*, Charles Singer (ed.), *Studies in the History and Method of Science*. Oxford: Clarendon Press, 1917, p. 237.
4. Edward O. Wilson, "Sociobiology: A New Approach . . . , p. 342.
5. Harry F. Harlow, *Learning To Love*. New York: Ballantine Books, 1971.
6. E. O. Wilson, "Sociobiology: A New Approach . . . ," p. 344.
7. *Ibid.*, p. 345.
8. Richard J. Barnet, *Roots of War*. Baltimore: Penguin Books, 1973, p. 5.
9. E. O. Wilson, "Sociobiology: A New Approach . . . ," p. 345.
10. Edward O. Wilson, *On Human Nature*. Cambridge: Harvard University Press, 1978, p. 19.
11. D. A. Lehrman, "A Critique of Konrad Lorenz's Theory of Instinctive Behavior." *Quarterly Review of Biology, 28* (1953): 337–63.
12. Jean Finot, *Race Prejudice*. New York: Dutton, 1907.
13. Alfred D. Low, *Jews in the Eyes of the Germans*. Philadelphia: Institute for the Study of Human Issues, 1979.

MARY MIDGLEY

RIVAL FATALISMS:
THE HOLLOWNESS
OF THE SOCIOBIOLOGY DEBATE

Historians in the future, if there are any, will of course study our age as we study ages already past, seeing what we ought to have done, and wondering how we came to make the mistakes that we do. As usual, they will find it hard to understand how we missed the clues which will be plain to them—clues which undoubtedly are already staring us in the face. So far, this story is a common one—a normal historical predicament. But in one way the present age is rather exceptional. There are at work today an unparalleled number of highly trained and capable scholars, with a wealth of information in their hands quite unthinkable in earlier times. I am not being cynical, nor underestimating our immense actual achievement when I ask, with this really splendid array of academic resources, could we, both on practical and theoretical questions, do a bit better? Is there an element of chronic waste in the system?

The reader will see that, following the general theme of this conference, I have my eye on over-specialization, on the walls which block knowledge from flowing where it is needed. But I want to start my contribution from a particular aspect of this evil —namely, academic feud and controversy.

Paper first delivered on December 1, 1979, at the Conference on Humanistic Disciplines in Transition, sponsored by The Society for the Humanities, Cornell University, Ithaca, N.Y.

Consider for a moment the lessons of the past. If we ask what
was occupying the best brains of any past epoch—what prevented
most of them from seeing the problems that they really ought to
have been tackling—we usually find that they were divided into
warring tribes by some vigorous but unprofitable dispute. Sometimes it was unprofitable because it was about something really
trivial—like the Biblical prophecies that preoccupied Newton.
Sometimes (more distressingly still) it was on a serious topic, but
was conducted as a fight between two half-truths, when it ought to
have been resolved by combining them. This happened with the
long dispute between Justification by Works and Justification by
Faith. And it still happens with the Free-Will controversy. This controversy has become entangled with our present theme, because
a number of people have argued, very strongly, that the acceptance of genetic causes of behavior involves fatalism, while the
acceptance of social causes does not. This is a position which could
only look plausible to somebody distracted by the heat of debate;
it makes nonsense of causality. I shall discuss it further in a
moment.

Now of course I am not condemning controversial pugnacity altogether. To do that would be to commit one of those false divisions into half-truths that I've just complained of. Of course
we need to be tough, sharp, and critical; sometimes we need to be
intolerant. There are limits to the value of tolerance. But what is
less well known and needs saying is that there are limits to the
value of intolerance too. Truth, even on small subjects and still
more on large ones, is seldom pure and never simple. As John
Milton said, the truth is never to be found in one piece, ready-made.
Instead, it has been cut into a thousand fragments and scattered to
the four winds. And ever since that time (Milton went on) "the
sad friends of Truth, such as durst appear, imitating the careful
search that Isis made for . . . Osiris, went up and down gathering up limb by limb still as they could find them. *We have not
yet found them all,* Lords and Commons, nor ever shall do, til her
Master's second coming. . . ." Milton was writing (in the *Areo-
pagitica*) for a limited purpose—simply to defend freedom of
speech. He was attacking the Puritan Commonwealth for its censorship of books. But censorship is only one expression of the in-

tolerant blindness that grows out of controversy. If intellect is to be properly used, we intellectuals in all ages need constantly to bear in mind the plea that Cromwell made to those same stiff-necked and confident Puritan sectarians: "I beseech you my brethren, in the bowels of Christ, think it possible that ye may be mistaken."

So how does this apply to the present controversy? As you see, I am going to suggest that we cannot deal with sociobiology on tribal lines. It is neither a heresy to be hunted down, nor a revealed doctrine necessary to academic salvation. It is instead the usual kind of mixed picnic hamper which needs to be unpacked, filled with the usual mixture of the nutritious and the uneatable, insights and mistakes, old and new material. Before opening it, however, I want to make a few more general remarks about the tribal approach, and especially about certain incitements to tribalism which are specially strong at the present day.

I begin, without apology, by a rather vast question. Is the point of academic enquiry to find right answers, or to avoid finding wrong ones? At a glance, these two enterprises look rather alike. Actually, they diverge surprisingly. And a lot of features in academic life direct us to the second rather than to the first. An academic is usually addressing an audience of fellow-experts who, in spite of their disagreements, do agree on a wide range of basic assumptions. If this audience is small, he can often find strength to resist them. But at a certain point the mass goes critical; he becomes incessantly conscious of their eyes on his back. Unless he is exceptionally tough, their expectations will dominate him. It will he hard for him even to think of taking a quite different line. So, how is any mistake which these experts are all making to be corrected? (If anyone doubts, at this point, that they *could* all be making such a mistake, I suggest that he looks up the back numbers of journals in this own subject for ten, twenty, forty, fifty years ago. . . .) This difficulty is a very serious one, because what restrains us here is not just our faults—laziness, timidity, and so on—but our professional virtues. It is our business to be fair-minded and humble—to avoid hasty bias, to bring evidence for all our views and evidence for that evidence, and to be prepared with further evidence for every bit of it. We are trained to fear rashness.

The upshot is that we tend to think of rigor as consisting simply in the careful practice of these defensive and negative techniques. It is really surprising how freely we use military metaphors of *attack* and *defense,* and how seldom, by comparison, we talk of a journey, of *advancing* toward the solution to a larger problem, or of using complementary tools or crafts in *building* the conceptual scheme that we need. We come to expect the information we gather to be used in consolidating a safe position rather than in making a new one possible. In the small hours, it begins to seem as if the worst thing that could ever happen to us in our professional lives would be to make a mistake, and have it exposed in public. There is not much to remind us of how much worse it would be to have entirely wasted those lives, and a great part of other people's too, in saying nothing. Among critics within our own discipline, we are only too conscious of Cromwell's warning: each of us constantly suspects that he may be mistaken. But the thought that those critics, whose ideas we have accepted, might all be mistaken becomes almost unbearable. Because individually we are so diffident, tribally we grow dogmatic. In this way the mere increased number of enquirers is counterproductive. The life-time of mistakes is prolonged. Academic isolationism sets in. We use compasses that orient us to our own ship and its crew, rather than to the ocean on which we are all sailing. We grow dedicated to security from error, as locally defined. This is fairly tragic, since of course that security is unattainable. We are all fallible. Everybody makes some mistakes. The exposer can get all of us in the end if he just keeps watching. The only place completely safe from exposure is the tomb. The only truly successful defensive rigor is *rigor mortis.*

Now of course, this account may seem something of a caricature. We all know that in principle we ought to say something positive, and if possible something big. But it is really hard to do so, and most of us do still normally expect that new thing to lie within the framework of assumptions at present being used in our discipline. There are occasional very interesting exceptions to this rule, academics who simply move out on their own and start a new way of thinking without giving a damn what anybody says about it—Wittgenstein, G. E. Moore, F. R. Leavis. To become such a person, however, you need, not just originality of mind, but also, and perhaps primarily, a hide like a rhinoceros. All these otherwise very diverse men had that advantage. Whatever their sensibilities on other matters, it simply never crossed their minds to worry about this sort of criticism. The result is rather curious. People like this are accepted as licensed prophets, and are quoted as authorities, but nobody ever thinks of imitating this feature of their method. At present, no philosophical article is complete without its quotation from Wittgenstein—("for what saith Wittgenstein?") This quotation is often brief, isolated, extreme, obscure, rhetorical, and dogmatic. The quoter, however, does not usually complain of this. Nor does he argue as follows: "Wittgenstein didn't bother to bring any support for his opinion, so I shan't bring any for mine; you can work it out for yourselves and be damned to you." *He* gives his references. The prophets don't influence general practice. The virtue of courage does not get cultivated. Most of us still continue to work within the existing local assumptions and to take them for granted. We don't expect to have to defend them, and we are not trained in doing so. Consequently, quite advanced specialists, including prophets, when confronted by unexpected criticism of these general assumptions, tend to flounder helplessly, like deep-sea fish brought suddenly to the surface. In the physical sciences, their indignation is usually silent, since they expect people outside their own subject to be fools anyway. In the social sciences and humanities, however, they are apt to explode in cries of inarticulate outrage, and to accuse their critics of unspeakable political motivations. Both reactions plainly show that they did not in the first place understand fully the assumptions on which their own method was based. All such deep assumptions lead beyond a single discipline, and are connected with those used for other topics and other forms of thought. No scholar ought to be surprised to find that somebody outside his specialty is in a position to criticize his assumptions, and can see clearly relevant points about them which he himself has never noticed. To think about these general assumptions is to see that they do have these outside connections, and to gain some idea of the direction from which trouble may come. Anyone who has considered this will be partly prepared for criticism and, when it comes, can soon set about seeing what it amounts to. With any luck, he may even have thought of it already for himself. He can get it in focus and deal with it as a limited problem, rather than as a general tribal threat which can only be met by war.

This is the skill that academic rigor requires of us. Unless we know where our own province stands on the wider map of knowledge, we cannot grasp properly the principles which apply within it. The supposed rigor of the isolationists is a fraud. A specialist with no grasp of controversial issues builds on sand. Rigor itself calls for some skill in general controversy, and in distinguishing the method appropriate to each stage of that controversy. As Aristotle put it, "it is the mark of an educated man to know just how much precision each enquiry admits of." This sort of education is difficult, and nobody is going to achieve it completely. People engaged on specialized work cannot be blamed if their first reaction to disturbance is one of irritation and confusion. When we have had time to recover from our first shock, however, we ought to be able to accept the issue put before us as a proper demand for attention to the wider map, to the communal road and water system, the set of assumptions which unites and serves us all. We ought not to harden our first mindless irritation into a feud. Instead, we need to set about overhauling the set of concepts that we have so far taken for granted. Hard? Of course it's hard. Real interdisciplinary work includes some of the hardest thinking in the trade, though the substitute activity that sometimes passes under its name includes some of the easiest. People manufacturing the substitute have only got to follow the well-known recipe for becoming an expert in Chinese metaphysics: always talk Chinese to metaphysicians and metaphysics to the Chinese; avoid short words, and never answer questions. People attempting the real thing may have to rethink their entire views from scratch.

When we do start on this work however, there are certain quite simple techniques that can help us. In the first place, we need to accept that every study does have underlying assumptions that are not yet clearly expressed, and which will only be made clear when they are questioned. We have to see that the general thinking needed to deal with these assumptions is not a layman's hobby, something beneath our scholarly dignity, but is an essential part of our own position. In the case of sociobiology, all parties have conspired to make the mistake of avoiding this general enquiry by underestimating the problem, by trying to treat it as a local issue, internal to their own subject. Both sides talk as if the very wide and fundamental assumptions involved could be proved or disproved by a few isolated experiments or observations of the kind with which they are familiar. This is like trying to change wheels with a darning-needle. These general assumptions are not particular facts at all, but commitments to whole ways of thinking, systems used for ordering and explaining large groups of facts. To "prove" or test them is to see how they relate to that whole range. Of course isolated facts that fit in badly with a new general suggestion need to be explained, but they can't constitute *disproof* of it. Every very general theory (including those fully accepted in the sciences) encounters some such awkward facts. Theories are compared with their competitors on their success and explanatory power across the whole field. In comparing them, we often find that theories that seemed incompatible actually don't compete; they are doing different explanatory jobs, and can be combined. This is the very common situation where there are two half-truths, and I suggest that where there is strong disagreement we ought always to assume that this is probably happening, we ought always to try it out as a first diagnosis, rather than deciding from the outset that those who disagree with us are either fools or knaves. We should look for reconciliation before we declare war. The inclusion of all fragments of the truth is our only proper aim. The "sociobiology debate" has been carried on with almost complete neglect of it.

So, perhaps there are two half-truths. What then? Well, our first business in such a case, if they do seem to compete, is to consider where the burden of proof lies. There is a clear principle to follow here. It is that the heavier burden of proof lies on the stronger and more extreme theory. *That* theory is the one which has the explaining to do. For example, in the recent row about admitting innate as well as environmental causes of human behavior, the extreme positions are those of people who want to exclude one of these sets of causes entirely. Plainly this sort of view is stronger and initially less probable than any which admits both sorts. Now, no serious writer among those who have lately drawn attention to innate causes has denied, or even attempted to minimize, the importance of environmental causes. All of them, from Lorenz to Wilson, have stressed the immense importance of culture, and have simply asked that *both* sets of causes should be attended to. It is quite true that the wilder rhetoric of sociobiology, the mystical language in which genes are sometimes treated as demonic fates, is inconsistent with this, and suggests a crazily extreme view of causality. This rhetoric must be considered, and I shall deal with it shortly. But the rules of good controversy demand that we concentrate chiefly on an opponent's official view, on his relatively sane and central position, not on aberrations which make no sense in any case. And Wilson leaves us in no doubt that, when he thinks about it, he fully recognizes the importance of cultural causes.

On the extreme environmentalist side, however, plenty of people in the social sciences *have* quite explicitly taken the position that innate causes could be ignored. Because this position is the stronger, it is they who have the arguing to do. Two things have obscured this point. First, the position was familiar to them, and familiar positions do not seem strong. We are used to them and don't ask what they are costing us in unnoticed error and confusion. Thus, the notion that there are no innate causes, which was introduced without any serious argument at all by the founders of behaviorism, has prospered unexamined for a number of pragmatic reasons to the point where, in spite of its total obscurity, many honest citizens believe that they cannot defend freedom without it. Secondly and in consequence, the absurd terms "genetic determinism" and "biological determinism" have been coined and applied indiscriminately to everybody believing in innate causes, with the implication that such people did not believe in causes of other kinds. This is simply a mistake. The notion of a number of competing kinds of hyphenated determinism—economic, physical, social, or genetic—is hopelessly obscure. Determinism is simply a general assumption that causes work, not a decision to back one set of them against another. Such backing is senseless; whole sets of causes cannot possibly be in competition like this. Marx didn't stop believing in physical or genetic causes when he drew attention to economic ones, nor do sociobiologists have to stop believing in social causes simply because they believe in genetic ones. The effect of this misconceived phrase has been to make the position of both ethologists and sociobiologists look much more extreme and bizarre than it is, and so to give the impression that the burden of proof lay on them. Many social scientists have, I think, now grasped this point, and see clearly that both extreme positions are untenable. Polarization is futile. We don't have to make an agonized choice between two alternative sets of causes. Instead, in each particular case we can simply work out *which* causes, from both sets, give us the best explanation. And the really alarming thing is not determinism but fatalism—an error both sides have been committing, though neither needs to.

A very interesting effect of this point about the importance of general assumptions is that one can easily be mistaken about where the borders of one's own competence lie. Instances of this catch the eye at once on both sides of the current debate. To start with sociobiology—Edward O. Wilson, when he complained (perhaps rightly) that the critical articles in the *New York Review of Books* and similar papers don't contain enough up-to-date science, went on to accuse them of using instead "historical anecdotes, diachronic collating of outdated, verbalized theories of human behavior, and judgements of current events according to personal ideology." (*On Human Nature,* p. 203.) Wilson clearly supposed that he was still on familiar ground here, and knew what he was talking about. As far as the omission or distortion of scientific evidence goes, he probably did, and it is a pity that he didn't go further and specify that part of his complaint. But after that he steps off into pure flapdoodle. What would a *non*-verbalized theory of human nature—or indeed of anything else—be like? Have sociobiological theories succeeded in becoming non-verbalized? How could you discuss history without using anecdotes? Or would it be a good thing for judgments of current events not to express the personal ideologies of the people making them? Wilson is here doing not biology but criticism, and because he is not interested in understanding it, he is doing it very badly. He has missed the opportunity to distinguish good criticism from bad by pointing to particular cases where the neglect of the sciences has actually been damaging. Instead, he has managed to commit himself to an absurdly destructive position, which makes all criticism, including his own, quite impossible. Very similar things happen when he discusses religion, making the cheerful suggestion that humanity's religious needs can all be met by scientific materialism, notably by a belief in the Theory of Evolution, because "Scientific materialism is the only mythology that can manufacture great goods from the sustained pursuit of pure knowledge." (*On Human Nature,* p. 207.) He is perfectly confident about this too, and does not seem to have any more suspicion that he might not know quite what he is talking about than did those sage characters who, in the French Revolution, enthroned the Goddess of Reason on the altar of Notre Dame and expected the populace to worship her. This blindness flows from his unshakeable conviction that science and religion are direct competitors, rival manufacturers who market, along with other contenders, competing versions of a single product, namely "a proper foundation for the social sciences" (p. 195). The loose metaphor "foundation" expresses and confirms this mistake. A building can have only one set of foundations. Just so, in the view of Wilson and many other specialists, a large phenomenon can, in the end, have only one real explanation. So if their own form of explanation has been shown to have some serious value, all others are thereby shown to fail or to be relatively superficial. But explanation is by no means such a simple, gravitational affair. To explain is to answer questions, and as many kinds of question can arise about a thing as there are reasons for being interested in it. The purposes, from which these questions arise, can of course be related, but not reduced to one another. There is no reason to expect that they will form a hierarchy toward a single "fundamental" set. The unfortunate thing about this kind of mistake is that it prevents its victim from expressing just what particular contribution his own kind of explanation does have to offer. Wilson is actually right in thinking that attention to evolution can in a special way improve our understanding of culture. But because he overstates his claim, this vital but relatively modest point is lost.

This narrow and competitive view of explanation is, as will soon become clear, one which Wilson shares with his opponents. Both parties express it in extraordinary language about causation— language that can only bear the sense that some causes (namely,

i. why not reduce Biology to
astro physics - the big bang?

On what methodological grd. is one claim ludicrous - Biology's & astro physics &
the other claim - Hu, S.S. Hist. Are Biological sust'ate Rational?

one's own set) are more real than others. When Wilson says that
"the organism is only DNA's way of making more DNA" or is
only "the vehicle" of genes, he uses a rhetoric that suggests that
evolutionary causes are the real, determining causes, while causes
which seem to operate during the life of an organism are actually
somehow idle. What sense can we make of this? One sense that
Wilson gives to it is a very simple one. He means that evolutionary
causes are the older—that because the long course of pre-cultural
evolution has in some sense determined all of more recent history,
inicluding the whole development of culture, it can also explain it.
He is right up to a certain point. The account we give of culture
must indeed be *compatible* with its evolutionary history, and must
make use of that history where it is relevant. But when relatively
simple causes produce more complex effects, they can't supply the
full explanation for them, which is one reason why not all explanation is causal. And of course the remoteness of causes by no
means increases their explanatory force. If it did, the original big
bang would be the only true explanation of everything, and we
all ought to be doing astro-physics, not evolutionary biology. The
concepts of cause and explanation, which are essential tools of Wilson's argument, are far subtler and trickier to use than he supposes.
He thinks he is quite at home in handling them—but he is wrong.

It is easy to find fault with Wilson. But considered simply as exploration his method is right, and I'm saying that we ought to be
more willing to go exploring. His expedition is in order—*provided*
that he is prepared to treat what he says as preliminary and to accept correction. When he finds that some doctrine which seems
crazy to him is being used in another discipline and is held to
discredit his views, he doesn't just sit at home and grumble, as
many scientists would do. He stumps off down the road toward the
alien phenomenon and does his best to shift the obstacles to understanding. When he can't move them, he hollers for the owner
to come and help him. Sometimes—as happens whenever he talks
about either language or religion—he is right off the road and
needs to be told so. Sometimes not. There is however also a more
general difficulty about map-reading. Wilson, like other physical
scientists, tends to expect all relations between disciplines to conform to the simple and elegant model provided by physics, chem-

ii. positivism

istry, and biochemistry. These subjects are indeed cozily linked in sequence, because they answer a group of closely related questions. The social sciences have often tried, with varying success, to pile themselves on top of this pyramid. They pay very heavily for doing this, and no sort of balancing act can prolong the process into the humanities. Logic and linguistics, history and law, geography and ethics, as well as much of biology, ask quite different kinds of questions. They slice into the world from different angles, and no sort of reduction will make them look like even bad imitations of physics. Wilson finds this quite incomprehensible, and this is why his "new synthesis" badly distorts the social sciences, and confronts the humanities either with shocked silence or with a bottle of rat poison.

Nonetheless, his attempt at mutual understanding is a proper one, and his demand for explanation is reasonable. On matters of general interest, we do all owe each other some explanation of our methods. Serious enquiries deserve an answer, even when they come from people who don't share our terminology. Wilson ought, if possible, to be met with genuine cooperation rather than territorial outrage. In spite of the imperialistic language he sometimes uses, he is not by temperament a Genghis Khan. Yet in spite of his personal virtues, sociobiology as a movement is a real menace, because it provides simple-minded people who like the jargon of science with an exceptionally slick set of catchwords and formulae for universal explanation. Like any flag-waving movement, as it gathers strength it is bound to collect a mass of supporters who will catch their leader's confidence without his scruples and without understanding his limitations. Because of the tradition I mentioned earlier, the academic world is full of people who ask nothing better than to settle into such an army. Wilson is a prophet, and he isn't going to lack acolytes. If we take the Wilson who appears most of the time in *Sociobiology* as the Dr. Jekyll of the movement, Hyde lurks in a few sections of it, emerges further in *On Human Nature,* and is fully unveiled in the cheap, crude, B-feature fatalism of Richard Dawkins's book *The Selfish Gene.* The cover of *Time* magazine's sociobiology number, showing a couple of lolling puppets making love while invisible genes twitch the strings above them, gives the level at which the message can be expected to reach the general public.

Acolytes are found, however, on all sides and not just behind Wilson. To balance the issue, I therefore turn next to a very similar error of judgment on the environmentalist side, namely, the failure of social scientists to notice that they have been doing biology, and doing it quite as badly as Wilson has been doing his criticism.

Those who used to take it for granted that behavior had no innate causes used to think of that belief as something safe and economical, and certainly as something internal to their own subject, not as a piece of biology. It seemed safe and economical because it was familiar. But with beliefs, as with plumbing systems, inspection often shows that the familiar is far from safe. In fact, because this thesis is such a strong and extreme one, the costs of supporting it by argument are extremely heavy, and the argument needed cannot be confined to the social sciences. To claim that people are infinitely adaptable—either generally or in some particular respect—is not to exclude biology from one's account. It is to claim that certain remarkable biological mechanisms exist which make that degree of adaptability possible. Inertness is not an absence of causal properties; it is a very special set of them. Now in all animals, but particularly in the "higher" ones, mechanisms do indeed exist to produce a limited and relative inertness, that is, to make the creature receptive to its surroundings and capable of learning. But this cannot be done by making it indefinitely passive and absorbent, like some kind of Kleenex. It is done by what is called "open programming." And, as Konrad Lorenz puts it,

> A genetic program of this kind contains several individual programs for the construction of various mechanisms, and therefore presupposes not less information than one single closed program, but far more information which must be genetically transmitted.

An example, at a simple level, is the ability to get used to a new landscape and find one's way about in it. More advanced examples are the inborn capacity that a normal baby has to acquire a language, and the capacity for play which it shares with a normal puppy or kitten. There are, however, very many such programs, differing enormously in form. Lorenz after describing some of them, continues,

Attempts are made time and again to treat all learning processes in terms of a single comprehensive theory. But what is here called "learning" is in fact an imaginary half-way house between [these various kinds of process].

(*Behind the Mirror,* pp. 81–82)

In short, learning is not simply a standard action of the environment on an organism, it is always some selection of the organism's own special activities, and this is not less true of complex human learning than of animal learning; it is more so. When people learn, complex patterns of response are, on the face of it, present as well as outside stimuli. If these patterns are to be overridden or ignored, very good reason has to be given. The hypothesis of complete adaptability must work to sustain itself, and must do so on alien territory. For instance, someone who seriously wanted to show that the whole range of human sexual behavior was produced by cultural conditioning alone would have a terrible lifetime's work in front of him, and at least half of that lifetime would have to be spent doing physiology in order to show that the physical causes which seemed to be at work were in some mysterious way actually idle, and did not affect the outcome. And though sex is a particularly glaring case, the same is true of other human motives and capacities.

No study is an island. We may all of us at any time, while doing what we think is our own work, find that we have stumbled head first into someone else's, and made instructive mistakes on a topic which we did not even know existed. In what sense, then, can we usefully put up frontiers at all? Now of course every study does have its own peculiar methods, and this point can be expressed by speaking of its autonomy. Marshall Sahlins, for instance, defending the frontier of social science against invasion by the study of individual motives, says he is asserting "the autonomy of culture and of the study of culture." Insofar as this means that culture must be studied by its own special methods (and not by ones assimilated, e.g. to those of physics), of course he is quite right. Autonomy here is simply opposed to reductionism. But he clearly means more than this. He means that the study of culture is entirely self-contained. He wants it to be independent in the

sense that it does not need, and does not have to consider, any outside assumptions about any subject other than culture. This must
also mean that its propositions cannot conflict with any from outside. Now it is evident that plenty of other causes do affect human
beings besides cultural ones; for instance, all anthropologists must
take account of climatic, geological, and medical causes. So this
view, as much as Wilson's, involves the assumption that among
these many kinds of causes, some are more real than others, that
there is some sort of reason for taking only one's own causal explanations seriously. Thus Sahlins says,

> Culture is properly understood as an intervention in nature,
> rather than the self-mediation of the latter through symbols. The
> biological givens, such as human mating behaviour, come into
> play *as instruments of the cultural project, not as its imperatives.*
> (*The Use and Abuse of Biology*, p. 63, my italics)

This talk of a "project" as an agent using instruments is every bit
as irrational and incoherent as Wilson's talk of DNA as an agent
using organisms. Sahlins has slipped, as readily as Wilson, into the
language of fatalism. He personifies his own favored cause as an
agent. This puts him in the position of the psychiatrist who, when
his patient claims to be Napoleon, retorts crossly, "How can you be?
I am Napoleon." The fatalistic jargon means nothing, and without
it, both claims to exclusiveness fall equally flat. Nobody is Napoleon. Both genes and cultural projects are elements in human
history. Neither has the slightest right to be treated as its puppet-
master. Neither can intelligibly be picked out as giving the ultimate key to its meaning.
   Why should anyone expect that there would be any such single
key? Well, there were (you will remember) six famous blind men
who went to see the Elephant. Having evidently read their Popper,
they made their investigation by a disinterested empirical method.
Each in turn walked up to the startled animal, and laid his hand
on the first part that he came across. In this way, they were able to
conclude (respectively) that an elephant is a wall, a tree, a snake,
a fan, a spear, and a rope. Had a Freudian been added to their
number, he could have concluded that an elephant is actually just

a rather large penis. They then went away to write their Ph.D.'s and to dispute the merits of their rival findings for the rest of their lives. Now we have for our present enquiry a subject considerably bigger and more complex than an elephant, namely, the whole of human motivation. It is so complex that no one can avoid constantly making mistakes about it, so that academics trained in the tradition I have described find it indescribably painful to contemplate. They are therefore most unwilling to look at it directly. Instead, the sophisticated observer prefers to peer at the reflection of it on the highly polished surface of an extremely distant set of events. Thus Perseus, unable to endure the sight of the Gorgon, turned his back on her and fought her by observing her reflection in his polished shield. Similarly, bandits in car-chases in films sometimes identify and even shoot their enemies in the car behind by studying their images in the driving-mirror. This is very clever. But it would, I think, be even cleverer to make firm discoveries about the motives which move us here and now by peering at the distorted images seen through the lens of population biology, reflected on the immense, remote, and shadowy depths of human evolution.

Exactly the same trouble however besets us when we try to study this subject solely by the methods of anthropology. Remote comparisons are useful—indeed they are absolutely necessary—as a corrective, but they cannot possibly supply the place of our primary method—hard thought, and good descriptive concepts, focused on the data immediately around us, on our own hearts and the people we really know. If sociobiology or anthropological arguments appear to show that something which is clearly taking place all around us cannot take place, then it is those arguments, and not the immediate data, which have to be thrown out. The confidence with which academics now accept such roundabout methods amazes me, since all of us know how quickly the fashions in such reasonings change and how riven with disagreement all such subjects are when we look into them at all closely. Thus Marvin Harris, in *Cannibals and Kings,* attempts to decide questions about the basic structure of human motives here and now, by using an incredibly elaborate and far-reaching argumentative path. His reasoning depends (among many other matters) on comparative data of the *average* heights of all people living 30,000 and 20,000 years ago; on full knowledge of the extent of protein shortage in all important cultures, and therefore of all the various alternative sources of protein available to them; on a full, detailed understanding of the practice of cannibalism among the ancient Aztecs; and on a correct appreciation of the relation of warfare to female infanticide among the Yąnomamö, and indeed everywhere else where either occurs. All this in a context where few anthropological conclusions remain undisputed by other anthropologists for more than a week, and where any tribe visited by two researchers in succession is at once the subject of debate. All this too, in a world where anthropologists are human—so that their contacts with the peoples they investigate do not by-pass the human bias and weakness that afflict direct, face-to-face investigation of human motives, but merely exploit a different, less well-understood area of it. (Anthropology, after all, is just another part of Western culture.) Harris, like Sahlins, works throughout on the crazy notion that in explaining a very complex thing like warfare, one must choose between entire alternative sets of causes—that there will be *either* an explanation in terms of human motive *or* one in terms of the economic factors he favors, not one which includes both. Amazingly, too, he considers his chosen alternative the less fatalistic of the two,

> War and sexism will cease to be practised when their productive, reproductive and ecological functions are fulfilled by less costly alternatives. Such alternatives now lie within our grasp for the first time in history. If we fail to make use of them, it will be the fault, not of our natures but of our intelligence and will.
>
> (*Cannibals and Kings,* p. 77)

But there are by now—and indeed always have been—all kinds of other economic investments in these and many other institutions. Anyone who takes economic factors, as such, to have a superior kind of force or reality, cannot simply select a few which fit his thesis. He must not confine himself to protein shortage; he must take on the entire economic scene, including the market forces which now obviously promote conflict. What Sahlins has done is do dodge the distressing suggestion that primitive peoples might actually *enjoy* war by testing his chosen set of economic factors as a closed, deterministic system up to the point where it suits him to abandon it, and then leap off the deterministic bus. He calls on us —our intelligence and wit—to change the institutions it has left us. But up to now we have appeared only as economic pawns. Who are *we?* And why should we find this change difficult? Our natures at this point must be taken into account. Do we—did our ancestors —actually view these institutions merely as means to an adequate supply of protein? Are there no other motives involved? Is the sale of war toys today, or the popularity of violent films, merely a race memory of protein shortage, or the inert, meaningless persistence of a dead convention? Have we really no direct way to answer this question? If not, why should we choose Harris's story rather than the dozens of others on offer? Economic fatalism is indeed a powerful contemporary myth, but its appeal lies chiefly in its tendency to make reform look impossible.

Now Harris has, of course, a reason for refusing to admit that dangerous nasty motives can be natural. He, like many other people, is convinced that if they were natural they could not be resisted. But this fear is quite groundless. It is a matter of constant daily experience, and a central fact in the natural history of our species, that people can to some extent control their bad motives. They are usually helped in doing so by the cultures which are another such central fact, but they can sometimes do it even without that advantage, because other, conflicting motives such as natural compassion intervene. Neither ethologists nor even sociobiologists in their saner moments have ever said anything to throw doubt on this obvious fact. Both parties have emphasized the number and complexity of our natural motives, so that the idea of giving any motive a walkover merely by admitting its existence is wholly contrary to their approach. On the other hand, bad motives certainly can be given a walkover by the self-deception that denies their existence. To admit that a weakness is real is not to increase it, but to put oneself in a position to resist it. Without knowledge of one's faults, self-control is impossible. With it, it is a normal condition of life.

But still, people may ask, how are we to know what our natural

motives are? Now when we want, in everyday life, to answer
questions about this, we bring into play a number of rich and
subtle conceptual schemes, which we are quite capable of using
to good effect. In academic controversy, however, many people apparently feel bound to ignore these schemes, and to treat any mention of them with contempt and embarrassment, as something
anecdotal and unscientific. Yet those schemes are the only possible
source of the concepts which we must use if we want to give substance to our projections. If we want to describe intelligibly what is
happening in remote cultures, or what we think happened in Pa-
laeolithic society, we have no choice but to use the everyday names
of acts and motives—theft and murder, spite and selfishness, affection, anger, ambition, and revenge. The suggestion that we cannot
identify these things directly is an extraordinary one. The peoples
studied by anthropologists do, it seems, sometimes express discreet
amusement at the weaknesses of Western man, and not least at
his occasional failures in the understanding of motive. But no
one has, I think, yet put it to them that this is no mere accident,
that Westerners are in fact a set of people who, unlike any other
group, simply possess no direct insight into such questions—who
have no grasp of motive, no idea why they do the things that they
have been doing all their lives, until they find out why people in
other cultures do them, and who prefer for this purpose to study
people as unlike themselves as possible. Were this made clear, their
mirth would, perhaps, sometimes get the better of their discretion.

On this point the record of sociobiology is, if anything, a little
worse than that of its opponents. Wilson has quite rightly been at-
tacked for the extraordinarily crude way in which he hitches everyday concepts, such as those that I have named, directly onto evolutionary genetics. Actually, this mistake seems to me so bad that it
may well completely cancel out the scientific advantages of his
method. As geneticists have pointed out, unless we know what a
single heritable unit of behavior is on the human scene, the whole
elaborate games-theory game will become unplayable. There is
something quite extraordinary here about Wilson's attitude to
ethology. He treats it as a humble forerunner of his method, a
relatively unscientific competitor which his space-age methods
have largely put out of date or reduced to subordinate status. This

is rather like saying that, now we have economics, we do not need to take history very seriously. Sociobiology looks "scientific" because it is quantitative, because it is conducted largely in mathematical formulae. But unless the conceptual groundwork that determines what is being counted and predicted is properly done, the calculations will be just hot air. Wilson's complaint about "outdated verbalized theories of human behavior" shows how little he understands the amount of work needed to describe things in suitable words—to find conceptual schemes subtle and flexible enough to make useful analysis possible. Ethologists like Lorenz and Eibl-Eibesfeldt have made this their business. Starting with the immense advantage of taking common language seriously, and with endlessly patient habits of observation, they make every effort to develop a suitable language while avoiding reductionism, and so do justice to the full complexity of human and animal life while putting it—as it must be put—in an evolutionary context. This is not a rival enterprise to the calculation of evolutionary probabilities but a necessary preliminary to it. And so subtly is it in fact now carried out that no excuse remains for anybody in the humanities and social sciences to evade the challenge of Darwin and treat social man as an isolated miracle. To say, like Sahlins, that "culture is an intervention in nature" is to out-Wilberforce Bishop Wilberforce; at least the nineteenth-century church did see God as a uniting explanation of both man and Nature. Such positions are indefensible, and those holding them ought by now to withdraw their attention from Wilson's faults—which, however fascinating, are of little lasting importance—and set their own house in order.

Behind all these mistakes on both sides there lies, I have been suggesting, the deeper one of tunnel vision, the belief that one kind of explanation necessarily excludes another. Exposing this belief to the air ought surely to make it instantly wither. Can anyone doubt that there are many kinds of question? Or suppose that, even if causes are our business, we can expect to deal with only one kind of cause? Or give a meaning to the suggestion that one kind is always fundamental? With a subject matter as complicated as human behavior, there is room and need for a hundred different conceptual schemes. They need not compete, and we can profitably use all of them—provided that we are prepared to take the trouble to see how they are related. If instead we put all our money on one—and particularly if we use it with the euphoric over-confidence that both Wilson and Harris show—we might as well give up our enquiry. Elephants are not to be caught in jam-jars.

Note

A word seems called for about the list that follows. Since I am serious in thinking that the "sociobiology debate" as currently conducted is highly confused, and "sociobiology" itself much less important than either its supporters or its enemies believe, I do not give further reading about it, apart from particulars of the books I have mentioned in this paper. My own estimate of its importance, and of its place in the much larger issues which should really command our attention, can be found in my book *Beast and Man,* especially in parts 2 and 3. My ideas on what we ought to be doing instead of fighting are also to some extent expressed in that book, and somewhat expanded in "The Notion of Instinct." "Gene-juggling" (pp. 108–34 in this volume) deals with the wilder excesses of genetic fatalism found in Dawkins's book *The Selfish Gene.* I am aware that this somewhat one-sided demolition job may lead some readers to think that I have finally come off the fence in the "Sociobiology Debate," and can now be relied on to stop thinking and fight, along with the rest of the regular guys. This would be wrong. The special dangers of fatalism supported by sociobiological rhetoric decided me finally, and somewhat reluctantly, to write something about it. But I still think the error of ignoring innate causes is more serious than anything called for by the basic program of sociobiology.

Other books. Ryle's *Dilemmas* gives the clearest and best short account I know of the relation between multiple non-competing explanations. It is mainly found in chapters v–vii, but the general approach to dilemmas expounded throughout is very nutritious, and there is a good chapter on Fatalism (ch ii, "It Was To Be").

Ernst Mayr's paper is a clear and classic account of "open programming." This is a key notion, needed if we are to understand

how innate capacities can contribute to human behavior without
thereby reducing it to machinery. How do the higher capacities,
which on anybody's view we have, work? How should we think
of them? As soon as we touch on this topic we catch sight of a
whole landscape of necessary distinctions and essential information
—a landscape both so vast and so neglected that the mind would
boggle if we did not fortunately have at hand the proper guide-
book: Konrad Lorenz's *Behind the Mirror.* At one-ninth of the
weight of *Sociobiology* and without any of its ballyhoo, this powerful little book really cuts to the heart of the problem of human
uniqueness. It gives us the tools for considering the human mind
in its place in nature, and that not just without diminishing it, but
with a much sharpened sense of what makes it special. This is
achieved in the first place by putting it in its context. Lorenz
sketches in, briefly but very clearly, the essential facts about the
vast range of cognitive mechanisms found at pre-human levels,
and the part that many of them play in the complex of human
faculties. By getting some grasp of the complexity and variety of
these elements, we begin to be able to see the startling effects that
are possible from a new combination of them, which is, as he
stresses, far more than the sum of its parts. We see how this might,
without further miracle, make possible the further advances of
speech and culture:

> I have described a series of processes that are all prerequisites of
> conceptual thought, and thus of the rise of man. My task was made
> the more difficult by the fact that these various processes are not
> of equal significance, and cannot just be linked together to pro-
> duce a superior pattern or system, as one can plait osiers to make
> wicker baskets, but are of varying degrees of importance. . . .
> The integration of two pre-existing and independent systems pro-
> duces a new one with entirely new and unforeseeable characteris-
> tics and functions. It is necessary fully to have understood this
> process of creation to appreciate the fundamental revolution of all
> life, brought about by the coming into existence of the human
> mind. . . . (*Behind the Mirror,* pp. 161 and 167).

This is one—and to my mind the most serious—of many contemporary attempts to break the controversial deadlock that has long
paralysed attempts to grasp the nature of the human thinking

subject. On the one hand, Rationalists have been prepared to take the topic seriously, but not to suppose that facts about the way in which minds (human and otherwise) actually work could be relevant to it. They proceeded abstractly, considering a thinking subject simply as a pure disembodied intellect. They thus drove an unbreakable wedge between the intellect and the rest of our nature, notably our feelings. On the other, Empiricists have been prepared in principle to consider facts, but not to entertain the notion of a thinking subject, lest it might turn out to be a Christian soul. Thus Empiricism demanded *a priori* a world consisting only of objects, and this was the root of its lasting conviction that minds must (in the teeth of all evidence) be treated as blank paper. In its last phase—dogmatic, metaphysical behaviorism—Empiricism became literally mindless, and it is in that unfortunate condition that it tries to grapple with the problems of human uniqueness in the pages of *Sociobiology*. But on the continent of Europe the dingdong battle between Rationalists and Empiricists has long ago been resolved into a number of less brutal debates, and the battle between the physical sciences and the humanities, which grew out of it in English-speaking countries, hardly exists at all. (It is not a law of nature that philosophers and biologists have to suspect and despise each other.) Lorenz, entirely unhampered by this load of prejudice and bad tradition, is at a great advantage, which he uses admirably. Eibl-Eibesfeldt, following this incomparably better method further with great imaginative sensibility and meticulous scholarship, is enormously illuminating. I cannot help suspecting that, if only his name were a little easier to pronounce and remember, we need never have had all this shouting about sociobiology at all.

BIBLIOGRAPHY

Dawkins, Richard. 1976 *The Selfish Gene.* New York: Oxford University Press.

Eibl-Eibesfeldt, Irenaeus. 1979 *The Biology of Peace and War.* New York: Viking Press.

——— 1979 "Human Ethology," A Survey, with replies to criticisms. *Behavioral and Brain Sciences,* Vol. 2, No. 1.

———— 1971 *Love and Hate.* New York: Holt, Rinehart & Winston; London: Methuen.

Harris, Marvin. 1978 *Cannibals and Kings.* New York: Random House; London: Fontana.

Lorenz, Konrad. 1977 *Behind the Mirror: A Search for a Natural History of Human Knowledge.* New York: Harcourt Brace Jovanovich; London: Methuen.

Mayr, Ernst. 1974 "Behavior Programs and Evolutionary Strategies," *American Scientist,* Vol. 62.

Midgley, Mary. 1978 *Beast and Man.* Ithaca, N.Y.: Cornell University Press.

———— 1979 "Gene-juggling," *Philosophy,* 54: 439–58.

———— 1979 "The Notion of Instinct," *Cornell Review,* Vol. 7, Fall issue.

———— (A collection of articles, title still undecided, is due out from the Harvester Press in the U.K., Cornell University Press in the U.S., in 1981. It will include "The Notion of Instinct.")

Ryle, Gilbert. 1954 *Dilemmas.* New York and Cambridge: Cambridge University Press.

Sahlins, Marshall. 1977 *The Use and Abuse of Biology.* Ann Arbor, Mich.: University of Michigan Press; London: Tavistock.

Wilson, Edward O. 1978 *On Human Nature.* Cambridge, Mass.: Harvard University Press.

———— 1975 *Sociobiology.* Cambridge, Mass.: Harvard University Press.

KARL PETER AND
NICHOLAS PETRYSZAK

SOCIOBIOLOGY VERSUS BIOSOCIOLOGY

The Problem

What is the interrelationship among man's biological nature, culture, and the environment? It is this very question which has been one of the major concerns of social theorists since the time of Hellenic Greece. With recent attention turning to sociobiology and ethology, this question once again has attained considerable importance. For the sociologist and anthropologist, it may be time to critically consider whether ethological and sociobiological theories can be of use to the social sciences in the search for the interdependencies between the nature of man and his culture.

At least since the beginning of the twentieth century, most sociologists and anthropologists in North America have maintained that man is "a social animal."[1] Man is perceived as living in a web of social relationships, dependent on others for support and responsive to the norms and standards of society.[2] Most sociologists agree with Broom and Selznick, who argue that human behavior is largely formed through social relationships amongst individuals acting together as members of larger groups.[3] Contemporary social theorists act on the belief, even if they do not always profess it, that man's inner nature is a *tabula rasa* which is fully dependent for its development on the processes of social interaction and socialization.[4] These processes together ensure the internalization of society's prescribed values and goals. Parsons, in this respect, has claimed that a sociological frame of reference need not take into account the interdependence of social action processes and the biological and physiological factors of their determination.[5]

Within the last fifteen years however, an extremely controversial set of theories has been introduced into the social scientific community. This body of theories is characterized by a sociobiological framework of analysis. Within this framework, a number of theorists have attempted to indicate that specific features of human behavior and culture are related to man's species-specific characteristics which have developed within an evolutionary context. The leading contributors to the development of sociobiological theory include among others, Edward O. Wilson, David Barash, and Pierre L. van den Berghe.[6] The question of the biological basis of human behavior is also central to the discipline of ethology. Ethologists such as Niko Tinbergen, Konrad Lorenz, and Irenaus Eibl-Eibesfeldt have endeavored to define man's innate nature in terms of "fixed patterns of behavior."[7] These innate forms of behavior are said to consist of sequences of coordinated motor actions which do not necessarily have to be learned and which are not mere reflexes.[8] Together, sociobiologists and ethologists have seriously questioned the liberal, social scientific belief in the idea of the social determination of human behavior.[9]

The Basic Principles of Sociobiology and Ethology

The leading advocates of sociobiology rely heavily on the theories of evolutionary biology in their interpretations of the social behavior of animals and man. A fundamental assumption that underlies at least the sociobiological theories recently formulated by Wilson, Barash, and van den Berghe, is that the behavior patterns of all living systems are adaptive in an evolutionary sense.[10] Considerations of "inclusive individual fitness" constitute the ultimate criteria by which the social behavior of animals and men are evaluated.

The central problem to which sociobiology is addressed has been succinctly stated by Edward O. Wilson as follows: "How can altruism, which by definition reduces personal fitness, possibly evolve by natural selection?"[11] The answer which Wilson and others have given to this problem is kin selection. Wilson maintains that "if the genes causing the altruism are shared by two organisms because of common descent, and if the altruistic act by one organism increases the joint contribution to the next generation, the propensity to altruism will spread through the gene pool."[12] Like other sociobiologists, he assumes that all biological organisms, including man, are compelled by the processes of evolution to maximize their "inclusive fitness."[13] According to this argument, the knowledge by the individual of his biological kinship ties is a necessary prerequisite for kinship selection to occur in human society. Wilson asserts that "the human mind [is] already sophisticated in the intuitive calculus of blood ties and proportionate altruism."[14]

The definition of culture which Wilson provides is directly related to his conception of "biological responses, from millisecond-quick biochemical reactions to gene substitutions requiring generations."[15] He views culture as an "hierarchical system of environmental tracking devices," insofar as "the rate of change in a particular set of cultural behaviors reflects the rate of change in the environmental features to which the behaviors are keyed."[16] While he is convinced that human culture evolves in conjunction with changes in the environment, he nevertheless sees culture and man's organic nature as separate. He emphasizes that "the specific details of culture are nongenetic, [in that] they can be decoupled from the biological system and arranged beside it."[17]

In a more recent work, *On Human Nature* (1978), Wilson attempts in one portion of the book to deal with the complex problem of the extent to which human cultural evolution is independent of man's biological evolution. To this end he cautiously indicates that "there is a limit . . . beyond which biological evolution will begin to pull cultural evolution back to itself."[18] While this point is well taken, it is not original. More importantly, what these biological boundaries actually consist of are never specified. In short, Wilson's views on the interactional dynamics which exist between human biological and cultural evolution are vague and unspecific.

David Barash, another leading sociobiologist, has recently indicated in his book *Sociobiology and Behavior* (1977) that living organisms, through the medium of kin selection, are able to "maximize their inclusive fitness—their net genetic representation in succeeding generations."[19] His idea of inclusive fitness refers to the summed consequences of both personal fitness (via offspring) and fitness derived via the representation of genes in relatives.[20] He feels that the chief task that sociobiologists face today is the application of the concept of inclusive fitness in the interpretation of human behavior and society.[21]

In dealing with the relationship between man's innate biological makeup and culture, Barash notes that human beings have an "unique dichotomous nature as both biological and cultural creatures."[22] It is further argued that "cultural evolution is very different from biological evolution."[23] He observes that "the whole range of characteristic disaffections with modern life may well have their roots in the dissonance between the relative complexity of our culturally mediated present."[24] He views the relationship between man's innate nature and culture as an antagonistic one.

Many of the insights and theories of sociobiology have recently been made use of in a specific sociological context by Pierre L. van den Berghe in his book *Man in Society* (1975). Unlike the sociobiologists however, he gives less emphasis to the relationships between human behavior and evolution. Instead, he attempts to depict what he feels to be the underlying biological bases of human behavior and social organization. Van den Berghe is in search of an answer to the question of "how man elaborated culturally in his biological propensities to aggression, hierarchy, and territoriality to create dominance-ordered hierarchies and political systems of extraordinary complexity."[25] He also emphasizes that man has an innate propensity for maintaining a family structure; religion; and play.[26] The interrelationships between man's innate nature and culture, as he understands it, is "the product of a complex interplay of biogenetical and environmental factors."[27]

Some ethologists have also tried to define the biological basis of human behavior. Ethology is the scientific study of the species-specific and genetically mediated behavioral patterns of animals and men. In dealing with behavioral homologies and phylogenies, ethologists have come to recognize the need for observing behavioral patterns, within the specific environment in which different behaviors have evolved.[28]

One of the leading European ethologists is Niko Tinbergen. In

Tinbergen's opinion, human behavior "is not qualitatively different from animals."[29] Underlying all human behavior, as he argues, there exist certain fundamental innate drives, such as aggression, sex, foodseeking, and the parental instincts.[30] These drives in turn are elicited by "innate releasing mechanisms" (IRM). He adds that certain human behaviors, such as scratching, sleeping, and sperm ejaculation, constitute innate displacement activities.[31] Group territorialism likewise is an innate feature of human nature.[32] In referring to the relationship between human nature and culture, Tinbergen has noted that "there are good grounds for the conclusion that man's limited behavioral adjustability has been outpaced by the culturally determined changes in his social environment, and that this is why man is now a misfit in his own society."[33] Like other sociobiologists and ethologists, he believes man's innate nature and his capacity for culture to be separate and even antagonistic processes.

Tinbergen's belief in the existence of innate drives and releasing mechanisms in man is shared by the ethologist, Konrad Lorenz. In *Studies in Animal and Human Behavior,* Vol. II (1971), he indicates that innate releasing mechanisms constitute fixed structural components of human society.[34] At the same time he argues that the processes of domestication, as well as culturally determined changes in human ecology, have resulted in the "dysfunction" of many innate species-preserving action patterns and response norms.[35]

In another work, *On Aggression* (1960), Lorenz observes that man's social instincts and his social inhibitions, "could not keep pace with the rapid developments of traditional culture, particularly material culture."[36] He asserts that man's innate killing inhibitions, as phylogenetically adapted behavioral mechanisms, are insufficient in controlling human aggression, in view of the invention of weapons utilized in modern warfare.[37] Cultural norms, as he believes, are necessary to keep phylogenetically adaptive behavior under control.[38] However, he admits that the exertion of "moral strength in order to curb . . . natural inclination into a semblance of normal social behavior" may result in nervous disorders and frustrations as experienced by the individual.[39] In this sense, man's innate behavior is said to be "phylogenetically so constructed, so calculated by evolution, as to need to be complemented by cultural tradition."[40]

Lorenz's *Evolution and Modification of Behavior* (1965) is an attempt in one respect to describe the interrelationship between innate behavior and learning.[41] He presented the argument that through deprivation experiments, applied to man as well as to other species, ethologists would be able to determine "where, in the species' system of actions, learning processes are biologically programmed, which are the reinforcements affecting them and how the unifying function of learning can be proved experimentally."[42] At least in this context Lorenz has defined the interrelationships between human nature and culture in a more unified context than many other sociobiologists and ethologists.

Eibl-Eibesfeldt, another well-known ethologist, has also tried, in a systematic fashion, to deal with the biological basis of human behavior. In one of his major works, *Love and Hate* (1972), he has contended that just because a behavioral pattern or disposition is inherited, does not imply "that it is not amenable to conditioning, nor must it be regarded as natural in the sense that it is still adaptive."[43] Human tendencies such as aggression are understood by him as being maladaptive, requiring some form of cultural control.[44] He defines man's tendency to cooperate within the group as being an innate human characteristic.[45] Eibl-Eibesfeldt is mainly concerned with the problem of how man is able to maintain his existence within individualized groups, and how the anonymous community which characterizes modern society produces problems of identification.[46]

Like Tinbergen and Lorenz, Eibl-Eibesfeldt is convinced that "drives, learning dispositions, and innate releasing mechanisms, can influence man's inclinations in a quite decisive way."[47] Culture, according to his point of view, is a means to control man's innate drives. Moreover, "the replacement of innate controls by cultural ones meant a gain in [human] adaptability."[48] Ritualized, innate behavioral patterns, as he points out, contribute toward the integration of the individual into the group and provide the groundwork upon which human cultures are constructed. These ritualized patterns of innate behavior include among others: rank and hierarchy, various sexual signals, greeting gestures and smiling, and the incest taboo. He feels that the readiness of the individual to form a bond with his fellows is based on an innate tendency of human nature.[49]

Sociobiology and Ethology—A Critique

The question of the applicability of the basic principles of sociobiology and ethology in coming to terms with the relationships between man's nature and culture has received wide attention from the social scientific community. Many social scientists remain convinced that sociobiology and ethology are unable to define what the biological implications of man's capacity for culture constitute.

Their main objection to sociobiological and ethological theories, therefore, is the alleged failure of these theories to describe the interdependency between man as a biological being and the effects of his cultural environment. Owing to the obvious complexity of this relationship, many sociologists and anthropologists insist that man's nature is unique and may not be compared with any other animal species. One critic, W. R. Bates, has emphasized that "the prime goals of biology should be the elucidation of those qualities that set the human being apart from the rest of creation."[50] He is convinced that past and present theories of comparative evolutionary biology have underestimated the uniqueness of the human being. Peter K. Smith, a British social scientist and a cautious critic of sociobiology and ethology, has observed that sociobiological theories "allow too much weight to the deterministic response of genetic change in natural selection and not enough to canalized genetic change greatly interacting with cultural processes."[51] The lack of a clear definition of the mediating processes among human nature, culture, and evolution is a much repeated criticism made in regard to sociobiology and ethology.[52]

A significant number of social theorists are convinced that the emphasis placed by sociobiology and ethology on the biological constraints of human nature lends support to the rather pessimistic idea that the human condition is unalterable.[53] For this reason, Edward A. Tiryakian has stated that sociobiologists like Wilson should make clear "what is the unique human genotype, . . . and just how social progress in the future can only be enhanced, not impeded by the deeper investigation of the genetic constraints of human nature."[54] From a slightly more radical frame of reference, Joseph S. Apler has suggested that sociobiologists have to take into account the fact that individuals, rather than being constrained by their innate nature, have the capacity "for changing their own nature, overcoming their environment and revolutionizing the institutions of society."[55]

Social scientists at present believe that man's behavior is mediated by social institutions. In the justification of this belief, David W. Paulson, as well as other critics, has condemned sociobiology and ethology for not providing "an account for example of the human evolutionary significance of role and status."[56] Given these various theoretical objections described above, the immediate impact of sociobiology on the methodology of the social sciences remains uncertain at this time.

Some sociologists and anthropologists, however, are convinced that sociobiology and ethology will play a positive role in the reorientation of the social sciences.[57] Despite the optimistic hopes of these social scientists, there are definite grounds to argue that neither sociobiology nor ethology has formulated an adequate interpretive framework, which could effectively explain the complex interdependencies among man's innate nature, culture, and the process of evolution. The failure of these various theories to deal with the fundamental complexities of these relationships becomes even clearer when their basic principles are subjected to critical scrutiny.

Probably the most significant shortcoming of Wilson's sociobiological approach, for example, is his failure to take notice of the fact that kinship selection within human societies is, in fact, mediated and defined by highly variable cultural criteria. This particular point was extensively argued in Marshall Sahlins's book, *The Use and Abuse of Biology—An Anthropological Critique of Sociobiology* (1977).[58] In reviewing Wilson's basic assumptions, Sahlins has provided a large amount of ethnographic evidence to support the argument that in many primitive societies "each kinship order has (accordingly) its own theory of heredity of shared substance, which is never the genetic theory of modern biology. . . . Such human conceptions of kinship may be so far from biology as to exclude all but a small fraction of a person's genealogical con-

nections from the category of close kin."[59] Owing to the theory that
kinship relations are culturally, rather than biologically, defined,
"the human systems ordering reproductive success have an entirely
different calculus than that predicated by kin selection and *sequitur
est,* by an egotistically conceived natural selection."[60]

At the same time, some sociobiologists have attempted to counter
Sahlins's arguments by emphasizing the role that reciprocal altruism
plays in human society. While the concept of kin selection does in
fact have a biological basis, reciprocal altruism does not. It is a
purely behavioralistic concept. To assume it will work where kin
selection would not, means to elevate reciprocal altruism to a human behavioral universal. In addition, sociobiologists have not provided a sufficient explanation as to why reciprocal altruism should
be a necessary condition of human social interaction.

Sahlins's argument, as presented here, places the basic theorems
of sociobiology in a questionable light. At the same time, Barash's
assertion that cultural and biological evolution are two separate
and autonomous processes contributes little to gaining an understanding of the interaction that has taken place among man's biological nature, his cultural environment, and the processes of evolution. On the other hand, as we have seen, Pierre L. van den
Berghe is acutely aware of the need for social scientists and sociobiologists to provide a clear conceptualization of what relationships
may exist between human nature and culture. His own insights into
the biological basis of aggression, the family, religion, dominance
orders, sexual dimorphism, play, and territoriality contribute to the
definition of the "biological structural" components of human behavior and social organization. However, these insights do not
provide a clear indication of what the dynamic processes of interaction are that take place between human nature and man's specific cultural situation.

The ethologists, even more so than the sociobiologists, have not
been able to elaborate a framework of analysis by which the dynamic relations between human nature and culture could be understood. Tinbergen's idea that human behavior is not qualitatively
different from that of many animal species, does not in itself explain the evolutionary or biological significance of man's cultural
capacities. Conversely, Konrad Lorenz gave slightly more attention

to the influence of culture on man's innate behavioral dispositions. However, he views this influence in terms of what he believes to be the dichotomy that exists between certain features of human culture and man's innate nature. Culture for Lorenz is merely a means to repress man's instinctual motivations, in the place of instinctual inhibitions that have been selected for during the course of human evolution. Finally, Eibl-Eibesfeldt has also done little to clarify the interdependent relationships among man's innate nature, culture, and evolution. Like Lorenz, he has simply pointed out that culture serves to control man's instinctual drives and inclinations. While he has pointed out that certain fixed behavioral patterns become part of the rituals of many cultures, he has not indicated how this ritualization process takes place.

The Need for a New Paradigm

Robin Fox and Usher Fleising optimistically stated that the essence of the sociobiological and ethological approach "is the acceptance of the synthetic theory of evolution as the master paradigm for the analysis of all the life processes, including such uniquely human processes as language and culture."[61] While we may sympathize with Fox's and Fleising's contention, it is quite evident that the ideal of sociobiology and ethology being the master paradigm for the analysis of all the life processes has not been realized to date. Neither of these biologically oriented disciplines has been able to generate an interpretive framework which could contribute to a clear understanding of the dynamic interaction that may take place among human nature, culture, and the forces of evolution.

For at least the last twenty years, biologists as well as social scientists have recognized the need to outline the ways in which man's innate nature and the dynamics of cultural evolution are mutually interdependent and complementary.[62] However, as a consequence of the dichotomy that many sociobiologists and ethologists continue to impose on the relationship between man's innate nature and culture, the study of human culture has been left to social scientists who lack any biological training.[63] At the same time, the belief held by social scientists in the cultural determination of human behavior "totally obscured the suggested complementarity of organic and cultural evolution."[64]

One of the major criticisms of the authors relates to the positivistic and reductionist view inherent in sociobiology. This positivistic position has satisfied those who call for a rigorous methodology although this methodology has not been forthcoming to date. The positivistic and reductionistic position of sociobiology creates the danger of refining the phenomenon to be studied—the human species—until it becomes unrecognizable from other organisms.

The analytical concepts of sociobiology are supposedly evolutionary factors not only outside the consciousness of the individuals to which few objections could be raised, but also beyond human influence—conscious, cultural, or otherwise. Such a position leaves out of consideration the evolutionary influence that human cultural creations have on human evolutionary development. But "human existence," as Kierkegaard said, "has a relationship to itself." Human evolution has a relationship to its own creation—culture. That is to say, a theory of human evolution must necessarily take into account the individual's incessant influence on himself or herself in the form of cultural creations, in addition to the modifications of the natural environment brought about by the culture that is created.

With culture in the picture, selective criteria such as "inclusive fitness" become untenable. The meaningful character of human cultural activities is irreducible to the motivation stemming from the net representation of a person's genes in subsequent generations. The reduction of altruism or reciprocity to the calculus of the genetic game is a denial that the human species influences itself through the great variety of cultural motivations, intentions, and strivings which may be either conscious or unconscious.

The essence of the human species is to act upon itself to become uniquely human. Human existence consists of an incessant act in which humans create themselves in whatever image they have of themselves. The conceptual and methodological framework of sociobiology has nothing to offer in approaching the evolution of *Homo sapiens* from this point of view. In the opinion of the authors, therefore, the positivistic limitations of sociobiology can only contribute to the investigation of a cultureless being—a being non-existent.

Owing to the fact that the human culture cannot be accommodated within the framework of sociobiology, many sociobiological theorists postulate therefore the parallel development of human biological and cultural development. This postulate, however, is not derived from empirical data, but stems from the necessity to safeguard the internal logic of sociobiology. Human biological development and culture cannot be allowed to interact. If they do, sociobiology dissolves into thin air. It is for this reason that sociobiologists insist on "separate" but "similar" processes of biological and cultural evolution.

Sociobiology attempts to explain the origins of culture by reference to their evolutionary concepts, not by investigating the culture creating capacities of the human species. Such a claim is analogous to explaining the origin of the steam locomotive by referring to the natural law of water expanding into steam when heat is applied to it. Whatever natural evolutionary forces produced *Homo sapiens* is one thing; once they became human and produced culture, they acted on themselves to remake themselves in the image of their own creation. It is this realization that necessitates the formulation of an alternative paradigm regarding the dynamic processes that exist among human nature, culture, and evolution.

An Alternative Paradigm of Human Nature, Culture and Evolution

The alternative paradigm to be outlined here closely follows the work of the German anthropologist, Arnold Gehlen, but is not identical with his theories. Gehlen is hardly known in the English-speaking world, nor were his major works ever translated. His monumental first work: *Der Mensch Seine Natur und Seine Stellung in der Welt* (Man, his nature and his position in the world) appeared in 1940 and went through eleven editions up to 1976.[65] His other writings include such monographs as *Studien zur Anthropologie und Soziologie* (Anthropological and Sociological Studies) (1963), *Moral und Hypermoral* (Morality and Hypermorality)

(1973), and *Urmensch und Spaetkultur* (Archaic man and modern culture) (1975).[66]

Konrad Lorenz in *Studies of Animal and Human Behavior* (1971) refers to Gehlen's key concepts frequently and accepts most of them. His discussion of these concepts, however, is of a piecemeal nature. Lorenz's failure to provide a systematic exposition of Gehlen's position is further confused by his attempt to equate some of Gehlen's concepts with his own. This is the case when he draws a parallel between Gehlen's notion of "instinct reduction" and his own concept of "domestication."[67] Gehlen rejects this identity. His research is concerned with the nature of man and its interrelationship with culture. He does not however develop an evolutionary framework of analysis.

Gehlen begins his inquiries into the nature of man by declaring two approaches to a formulation of the nature of man as being inadequate. (Note that the term "man" is used in the generic sense and is meant to refer equally to both men and women.) These approaches maintain the view first that man is nothing but an advanced ape and second, that the nature of man can be approached through an investigation of one or several psychological or physiological human traits, such as reason, mind, erect posture and language, and so on. Every single human trait can be found among animals in one form or another. Even creative intelligence and language can be found in primates as experiments with chimpanzees indicate. These frameworks of analysis are inadequate because they fail to account for the phenomenon of higher human mental life and its contents. Nor, as Gehlen points out, do these approaches provide insights into such questions as Why do human beings take cognizance of some but not of other things? What is morality and why does it exist? What is fantasy? What is human will?

Gehlen insists that an adequate view of human nature must be capable of answering these questions. He suggests that this is possible only if one frees oneself from the habit of assigning man a place in the zoological classification of the animal world. Nature, he says, has assigned man a special position. It has realized in man a developmental principle that never before existed in nature nor was ever tried by nature. The nature of man reveals a new organizational systemic which is decidedly different from anything found in animals. A biological anthropology, or better said, a bio-cultural view of man, cannot proceed from a contextual framework developed in zoology and biology, but must be derived from a study of man and man alone. This by no means implies that the theories of zoology and biology are useless but that they cannot be a starting point for the study of man.

He maintains that a bio-cultural concept of man must utilize an holistic perspective on man. Any causal reasoning from one trait to another, such as whether language caused intelligence or vice versa misses the point. Man, he asserts, is an unique holistic design of nature and man's survival capacity consists of a number of essential human components standing in mutual relations to each other. What Gehlen calls the bio-cultural nature of man is a system of human traits in interaction, whereby the existence of one or the other of these traits is predicated by the existence of all the others. Any change in one or the other of these traits requires adjustment on the part of all the other traits. He follows essentially a General System perspective long before General Systems Theory found its formulation through people like von Bertalanffy, Boulding, and Wiener.[68]

The common mistake made by anthropologists and biologists alike, is to accept Darwin's claim that the increase in the learning ability in vertebrates runs parallel to the zoological development of the species and reaches a peak in the development of man. Gehlen quotes Buytendijk who demonstrated that hunting animals, regardless of whether they are insects, crabs, birds, or mammals, have the same general instincts. However, categories like hunting animals, herd animals, tree animals, and so on, are performance categories, not zoological ones. Animals belonging to the same performance category show a high degree of similarity in their instinctual equipment, as well as in their learning ability. This is true for such evolutionarily different tree animals as apes, squirrels, and parrots. Squirrels have the ability to recover hidden objects according to visual memory data capable of reaching them via detours. These abilities are often seen as one of the achievements characterizing those on the higher level of zoological classification, like apes. Dogs on the other hand are incapable of doing so. Animals that are zoologically closely related, like the frog and the toad, show very different instinctual and learning abilities. The frog is a lurking animal, while the toad is a prey-seeking hunting animal.

In explaining the differences between animals and humans, Gehlen first outlined the systemic conditions under which an animal enlarges its (non-genetic) behavorial repertoire through learning in such a way that it increases its survival capacity. These conditions consist of the following:

1. With the exception of some social insects, every animal uses successful experiences only under the influence of vital stimuli. This means that a performance surplus for animals is only possible within the confines of concrete situations containing vital elements of attraction or avoidance.

2. Animals learn only in concrete present situations and use the experiences only in those situations that contain a release mechanism through which the learned behavior is discharged.

3. Every animal behavior is instinct related. It follows an instinct totally as in the case of a migratory bird that behaves in the direction of some future condition (summer in the south) of which the animal can have no previous experience or . . . its behavior is instinct related in the sense that it builds up learned behavior within the immediate field of its instinct governed, vital life interests.

4. All animal learning is strictly limited. What an animal learns, or more importantly, what it fails to learn, is circumscribed by its species-specific instincts and the environment to which they correspond. Within these limits, an animal is able to build up and expand its behavioral repertoire but it cannot transcend it.

The "fitness" of an animal, therefore, is determined by the systemic integration of its instincts, its learning ability, its senses and organs, and the conditions of its natural environment. Said in a different way, an animal responds to the environment in such a way that it recognizes those aspects of the environment that correspond to its vital life interests which are represented in its instinctual and physiological makeup.

Gehlen emphasizes that man is unique in that he does not fit into this animal design. Man is not a creature that is definable by natural fitness between organism and environment. The world he recognizes and responds to does not stand in the same relation of fitness to his organic and mental constitution, as is the case of animals.

As Gehlen claims, man is an "undefined creature" in the sense that he lacks the animal fitness between organism and environment. Because man lacks fitness, he is a creature who poses a problem to himself; a creature whose survival depends on the position that he takes towards himself. Man is an acting creature who is forced to translate the deficiencies of his "undefined nature" into activities that make survival possible. However, man is also a "jeopardized" creature; a creature with a high constitutional possibility of failure. Constitutionally, he is an organism with a high degree of improbability whose very existence depends on how successfully he is able to transform the improbability of his existence into chances of survival through his own actions. Man's evolution is a process of maximizing the improbable. Finally, man is a creature of foresight. He lives, in contrast to the animal, for the future and not in the present. Man, like animals, is a creature of performance. The type of performance demanded of him however, is qualitatively different from animals. The organization of man's abilities by which this performance is forced out of him required a new natural design. Let us turn to the evidence that Gehlen presents in support of the above characterization of man and the unique organizational principle of human behavior.

Gehlen points out that the components of the innate nature of animals consists of instincts, limited learning abilities, physiological structure, and the species-specific ecological niche (environment). These components stand in a dynamic relationship to each other. It is this dynamic relationship that constitutes the bio-structural determinant of animal behavior. None of these bio-structural components, nor their dynamic relationship is characteristic of man.

The instinctual equipment of human beings is characterized by an instinct reduction. What might have been the original instincts of *Homo sapiens* have become separated from any behavioral motoric. They are subject to inhibitions, that is to say, they can be kept internalized and do not necessarily lead to any actions. In addition, these instinct reductions, in contrast to the specific instincts of animals, are non-specific in the sense that they only develop into specific drives and interests in conjunction with culturally defined experience and learning. They become organized by the cultural and environmental stimuli. Through culture, human instincts become set in symbols and fantasies that form the context in which they are recognized and memorized. Retained in the mind in terms of these symbols and fantasies, they obtain rational recognition relative to specific culturally defined needs and interests. Finally, human instincts are variable. They are capable of following experiences and learning and grow with them. Lacking such learning, they may not develop at all. Human instincts are diffuse and can be joined or merged. In the individual there is no sharp division between elementary needs and derived needs. In fact, the term "instinct," as it refers to the constitution of animals, is not an applicable concept in reference to man, given the above mentioned characteristics. The term "instinct residue" will be used throughout the remainder of this discussion when referring to the nature of human behavior. This term is understood to refer to all of the points made above.

Turning to man's physiological constitution, in contrast to the intrinsic ecological fitness of animals, man as a world-open creature has characteristics which, in the biological sense, appear as non-specializations or primitivisms. Man has no natural protection against climatic change. Nor does he have natural tools of attack or any specialization for flight. His dental equipment is neither that of a carnivore nor that of an herbivore. In comparison with some primates who are highly specialized, man as a natural creature is deficient. Man requires an extremely long period of physical and mental maturation, longer than for any other animal. Under natural conditions, such a creature would be facing specialized escape animals and dangerous predators, making it highly probable that this creature would be either eaten alive or die of starvation.

Finally, man does not occupy any specific ecological niche. He is not fitted into a compatible non-exchangeable environment, congruent with his organic structure. Owing to the fact that man is physiologically a deficient creature, there is no natural environment that fits his morphological and instinctual constitution. Unlike animals who recognize and respond to a range of stimuli congruent with their vital life interests and thereby ignore others, man is subject to overstimulation brought about by his openness to the environment. Man is a world-open creature. He does not live in an environment of instinctually defined interpretations, but in a world of endless, unforeseen and startling surprises. To live in this overstimulating, unforeseeable, and dangerous world requires an organization of perception and responses.

The lack of concrete instincts, in combination with man's physiological deficiencies and his openness to the total environment, makes for a jeopardized, endangered creature whose first and foremost task is to transform the deficiencies of his existence into chances of survival. *Man must unburden himself from the lack of natural fitness, which characterizes animal existence, and ensure his survival through his own actions. These actions are systematized in culture.* These two sentences are the key to the understanding of the structural law which underlies the whole buildup of human performance.

Man's cultural achievements are to be considered from two points of view. First, they are productive activities to unburden himself from his natural deficiencies. Secondly, they are achievements that man must "force out of himself" which, in contrast to animals, constitute an entirely new mode of existence. Seen from this point of view, man's instinct reduction is as much a necessary prerequisite as is the emergence of mind, his physiological deficiencies, and his openness to the environment in the struggle for cultural organization.

The origins of human culture must be understood in relationship to the overall evolution of man. Most evidence of primate evolution is obtained from the Miocene epoch of the Tertiary which is said to have lasted from 12 million years ago to about 2 million. Higher primates during this important evolutionary period were able to achieve a level of adaptation that enabled them to significantly increase in both their numbers and range of territory. Different primate forms in this period successfully penetrated zones of habitation of a savanna type, which had previously been outside the range of the tropically adapted species which made up the primate order at this time. Whether this shift in habitation was due to population pressure, the shortage of foodstuff, or the expulsion of weaker and less adapted primate forms by stronger and better adapted ones remains an open question. Whatever the reason, this period of expansion or what is referred to as "adaptive radiation," led many primates to occupy ecological niches where they encountered different environmental demands and faced new and unprecedented selective pressures. These differing selective pressures played a significant role in the further evolution of these primate forms towards the development of the hominid type.

The origins of early man then, are directly linked to the transformation which some primates underwent from an arboreal existence to a terrestrial one. It is to these terrestrial primates that we owe our advanced conceptual sensitivity, locomotor ability, manual dexterity, and large brain capacity. In fact, the expansive flexibility of the human hand is the direct result of man's arboreal ancestors who left the trees and assumed an upright posture. Probably at first, similar to modern apes, these early hominid types walked upright only part of the time. Once bipedal locomotion became a permanent form of movement, the hands were freed from their locomotor functions and were able to develop abilities such as the use and manipulation of tools. This facilitated in turn the further evolution of the species.

This is not to say that the original arboreal primates did not make at least some use of their hands. However, the use of their hands by these primates was limited to such things as the building of nests, the acquisition of foodstuffs, and climbing. Conversely, the early hominid's superior brain, which controlled a much more flexible hand, made it possible to invent relatively complex tools and weapons. Man's dexterity with his hands also facilitated the emergence of writing and artistic expression. In addition, early man's acquisition of an upright posture, complemented with the gaining of manipulative skills in the use of his hands, freed the mouth from such tasks as fighting and tearing food apart. Liberated from these cruder functions, man's mouth, jaws, and structures of the throat evolved into a set of organs adaptive for the evolution of speech. Early man's large brain capacity further ensured the development of speech and language through its ability to manipulate the physiological speech mechanisms as well as being able to formulate the messages to be communicated.

The transition of the advanced primates from an arboreal to a terrestrial existence during the Miocene epoch was of course a gradual one. It is important to emphasize that it was not just their terrestrial adaptation that had a selective advantage for the evolution of man. The adaptiveness of these prehominids was also related to the fact that they were highly socially oriented. Their sociability would in itself be of a high adaptive advantage in competing for survival on the African plains where early man probably originated. Culture in itself must be seen as a means to organize and communicate these essential elements of human sociality. Insofar as the basic elements of culture consist of symbolization, communication, and tool use, it is obvious that the origins of culture are directly related to the lack of human physiological specialization, as explained above.

From this perspective it may be understood that there exists a profound contrast between the animal and human evolutionary processes, which reveals itself distinctly in the following short formula:

> The principle of animal evolution is compulsory adaptation to the environment by means of the body—*the principle of body compulsion.* The principle of human evolution is that of freeing man from the compulsion of body-adaptation by means of artificial tools —*the principle of body liberation.*[69]

The fact that the principle of body liberation constitutes the dynamic element of human evolution is evidenced in the progressive development of all those organs which were pre-eminently engaged in the employment of tools. For example, the human hand when compared to the long curved hand of the orang or chimpanzee is wider, shorter, and has a well-developed thumb. Together, all these features are indicative of man's physiological adjustment to the progressive development of tool use. At the same time the structure of the human hand also demonstrates the loss of man's physical adaptation for climbing in the trees like a monkey or ape. Thus, man's loss of body adaptation was closely associated with his gaining of body liberation through tool use. In turn, tool use provided the basis for material culture.

Evolution has promoted the physiological adaptation of animal species to the environment through the principle of body compulsion. In the case of man, evolution has facilitated the adaptation of the human body to the requirements of tool use through the principle of body liberation. The origins of culture, then, are directly related to the behavioral and physiological capacities which the principle of body liberation provided to the human species.

From a strictly Darwinian perspective, natural selection within animal species consists of selection on the basis of physiological adaptiveness. Conversely, within human populations, selection has consisted of promoting the survival of those individuals who have a higher degree of cultural capacity and inventiveness. For example, the evolutionary success of the physically inferior and weaker Cro-Magnon as opposed to the Neanderthal, is a demonstration that the adaptive survival of human populations was dependent on their possession of superior technologies and cultural abilities rather than on greater bodily strength. Thus, if cultural inventiveness was a key to human species survival, it should not be surprising that man's reliance on instinctually based behaviors was significantly reduced and diffused during the course of human evolution.

The instinct reduction which separated human behavior from instincts and led to a deconcretization of their focus and direction, created an hiatus that simultaneously made it possible for the human mind to develop. The reduction of instincts removed the certainty of the future existence of the species. At the same time, individuals were forced to rely on their own actions to create the chances for their survival, which in turn made it necessary to develop a mental structure capable of functioning in the direction of recognizance and foresight. It is for a good reason that human instinct residues are subject to conscious recognition and interference of the mind. These instinct residues must be conscious and flexible in order to be endowed with culturally invented symbols, fulfillment situations, and social relations. Only a creature that is conscious of its instinct residues is able to inhibit or displace them.

Elaborating somewhat on Gehlen's observations, such instinct residues or innate predispositions exist in three broad categories. There are first the organ related instinct residues, such as hunger, sex, speech (not language), and so on. Secondly, there are those that make individual human cultural growth possible. These range from the ability to take the role of others toward one's self (e.g., the development of the human self) to those that make human self-assertiveness possible. Finally, there are those broad dispositions that bring human group behavior within the range of high probability. Depending on their cultural orientation and symbolization, these can take the forms of group loyalty, group cooperation, and group aggression.

It is important to emphasize that the above definition of man's instinctual residues has been verified by a significant number of theoretical and empirical studies conducted in the areas of human psychology, evolutionary biology, sociology, and anthropology. The categorization of the organ related instinctual residues in man has been given extensive attention by psychological and biological theorists.[70] There is also empirical evidence to show that the instinctual residue of individual self-development and maturation and the ability of the individual to take on the attitude of others are in fact innate characteristics of the human species.[71] One of the strongest sources of evidence has been provided by Erik Erikson, in his theory of the life cycle and the epigenesis of identity.[72]

Erikson, on the basis of his own clinical observations, maintains that the maturation of the human being is characterized by a basic program of development or epigenetic principle, which is a species-specific characteristic of man. This epigenetic principle may be summarized as being the innate general plan of the personality from which the parts of the human self arise. This biological principle of self-development is comprised of eight sequential stages. In the order of their emergence in the personality, these stages consist of the development of the senses: trust, autonomy, initiative, industry, identity, intimacy, generativity, and integrity.[73] The development of each of these stages may in turn be facilitated or inhibited by the social environment in which the individual finds himself. Probably the most critical stage, according to Erikson, is the fifth, where the child begins to consider the opinions of others and develops role expectations, which in turn facilitate the emergence of a sense of self.[74] This epigenetic stage constitutes a species-specific characteristic of man.

Apart from the organ related instinctual residues and those related to the development of the human self, most sociologists and

anthropologists contend that the tendency toward group membership is a universal feature of the human species. This strongly
suggests that the capacity for group membership is based on an
instinctual residue.[75]

The instinct residues are an essential element of the definition of
humans as a unique species separate from all other animal species.
As species-specific traits, instinct residues are fully complemented
by the species-specific emergence of mind, which in combination
allowed humans to develop culture. All these processes are seen
in relation to the physiological and morphological non-specializa-
tion of *Homo sapiens* and their openness to all natural environments in which they have shown to be able to survive. While human nature and its primary components may be defined at a species
level, such a definition is static. To see the human species in an
evolutionary context is to inquire into the feedback relationships
among its biological nature, its cultural creations, and the specific
(or non-specific) environment in which its members choose to live.

To elicit these dynamic relationships one must first ask the question, what is variable in human nature? What are the differences
among the various individuals of the species; differences which have
a genetic base? Sociobiology and ethology want us to believe that
it is behavior that is genetically determined and therefore can be
extinguished, created, or proliferated through differential reproduction and mutation. In contrast, our assertion is that behavior is
an emergent product of human nature, culture, and the environment. Referring to human behavior as an emergent product simply
means that human behavior does not consist of the sum total of
biological, cultural, and environmental influences, where each type
of influence can be neatly separated from one another and analyzed.
Rather, it implies that human behavior is the product of the dynamic interaction among these various spheres of influence.

At the same time, human beings differ from each other and
some of the differences, it seems, have a genetic base while others
are a product of the interaction of genetic components with culture and a third are solely a product of culture. Our discussion of
instinct residues on the species level defined the structure of man's
drives as the result of an interaction of human genetic and cultural
predispositions; that is to say, the result of human constitutional

predisposition that is subject to cultural development. But for those of us who reject the *tabula rasa* doctrine, the difficult question arises that, if human drives are not genetically based but the results of genetic predispositions and culture, what is it that distinguishes one human being from another and has a genetic base?

The work of some members of the German Psychological Association during the 1920's conceptualized this genetic base of human individuation.[76] According to these theories, the human infant at the moment of birth or even before, faces the world not as a *tabula rasa* but responds to his or her environment through a definite set of genetically based primary functions.[77] The concept *primary functions* refers to an incomplete list of functions such as the level and type of individual attentiveness, energy, perseverance, and emotional addressability, which exist in the infant prior to cultural and environmental experiences. These functions allow him or her to face, internalize, and manipulate these experiences in terms of his or her own development. These individually different and genetically based vital functions inject dynamic qualities into the species-specific processes between human instinct residues and their cultural formation; giving these processes certain functional qualities but not determining the nature of these processes in the behavioral sense. A certain type of perseverance, for example, might be an advantageous element in a number of different cultural activities as so might be the lack of it. Whether such a quality is of advantage to the individual depends very much on how it is integrated with the rest of his or her species-specific biological nature, culture, and the environment leading to behavior that is the result of the integration of all these factors.[78] Human genetics consists of shuffling and reshuffling primary functions into new combinations in individuals, commensurate with the cultural development and the environmental concern of human groups and aiming at behavior which makes survival of the group possible. This is also the definition of human evolution.

One important consequence of man's instinct reduction, according to Gehlen, is the disposition of free excess motive power. The amount of motive power necessary for man to perform the vital functions for his survival varies greatly from one culture to another. Animal motive power is fitted to both its instincts and the environment, following a natural rhythm in which it is released and acted upon. Such instincts as sex, migration, and nest building are innate releasers of motive power in conjunction with certain conditions of the environment. Conversely, after the animal instincts have been satisfied, the motive power for its particular performance is withdrawn and the instinct is temporarily laid to rest.

In man, not even the organ related instinct residues follow any rhythmic environmental cues. All human instinct residues are addressable at any particular time. Such a condition presupposes the existence of a constant reservoir of free motive power which can flexibly be withdrawn whenever man faces adverse circumstances.

An instinct rhythm in man, triggered by environmental influences, would have disastrous consequences. His motive power needs to be oriented toward his cultural achievements, but cultural achievements do not uniformly follow any environmental cues. For the sake of the permanency of human existence, a ready pool of motive power is necessary. This pool is capable of enlarging itself in the process of its activation and discharge. Excess motive power exists *a priori* to cultural and social development. The long time period from birth to sexual maturation, Gehlen argues, is characterized by motive power which is largely free of sexual desires. That is to say, it is unattached to any concrete sexual symbols that might activate its discharge. The existence of this free motive power in the child is channelled into playful exploration of the social and physical environment. In the course of these activities, the child organizes and systematizes the world of external stimuli in terms of its own personality development. Only when this process reaches maturity and when language, thought, freedom of movement, and the perfection of motor skills are completed, does sexual maturity appear. It is at this time that the unattached motive power (which was necessary for the individual cultural growth of the child) is now partially drained into the specific area of sex.

The human motive power which was made available through the reduction of instincts is subject to the cultural influences and societal pressures encountered during the personal development of the individual. These pressures are utilized in the sense that culture provides the symbols on which the motive power orients and organizes itself. These orientations may inhibit instinct residues or they may facilitate them.

Finally, it must not be overlooked that the individual human being guides and directs this process in terms of his or her own unique organization at every stage of his or her personal development. Structural pressures in the transformation of motive power may be biological, as in the maturation processes mentioned above, or they may be cultural, as in deferred gratification. Regardless of their origin however, they are always purposeful in making it possible for man to maximize his self-acting capacity. *Self-acting is the peculiar human activity that makes survival possible for a creature that lacks the animal fitness and whose survival depends on the type of action that it is able to generate out of itself.*

Gehlen's characterization of the nature and role of human consciousness must further be considered. Consciousness is that entity which fills the hiatus created by instinct reduction. It is the vital tool that helps to orient and to endow human instinctual residues with symbols. Consciousness also has some remarkable limitations. These limitations are indicative of how it fits into the bio-cultural framework of man as developed by Gehlen. All vital biological processes of man, for example, proceed outside of human consciousness. We know as little about how we breathe or digest as we know about what we have to do to lift an arm. Nor is the inner world of consciousness itself accessible to the awareness of the human mind. Man lacks the sensory equipment that provides him with self-knowledge about his biological and mental functions. Instead, these functions are coordinated and directed by more reliable biological control structures.

The capacity of the human mind is exclusively directed toward the structuring of human behavior. It penetrates the totality of human existence only to the extent that it enables man to build up, direct, control, and coordinate those forms of human achievement that are necessitated by his life interests. Due to the variety and complexity of these life interests, an extensive variety of mind functions is necessary to translate them into such action patterns as language, memory, fantasy, the whole inner range of mental life of man and his powers of reflecting on the environment and on himself. The mind of man therefore stands in an inseparable relationship to his physiological constitution, to the "world open" environment in which he lives and to the culture which he creates.

The reduced human instincts and the capacity of the human

mind also stand in a dialectical relationship to each other. The reduction of the first made the growth of the second possible. The
relationship between these two components, therefore, is structured.
The diffusion of instinct residues and their capacity to join or merge
in combination with their ability for permanent stimulation led to
a constant competition among these residues to structure the field
of human action organized by the mind. Culture in this sense is an
emergent product of these residues and the mind. In man, the permanent ability to be sexually stimulated led to a sexualizing of
other instinct residues. Conversely, sexual residues are joined and
merged with others. It is owing to the simultaneous penetration
of instinct residues that personality conflict and ambivalence are
probable. On the other hand, it is exactly this diffuseness of instinct
residues that opens the possibility for stable human interests in the
form of tension structures.

It is possible that residues of submission and aggression synthesize into a stabilized tension whereby one residue becomes inhibited by another. The result is a new emotional state, a mental
tension of paying attention to an object to which the viewer stands
in an ambiguous position of mixed submission and aggression. This
stable tension represents a mental orientation toward the object,
that sensitizes the individual toward the characteristics of the object
and leads him to discover previously unrecognized aspects of the
object. At the same time, this stabilized tension allows for the build
up of an action pattern through the utilization of these newly perceived aspects and in so doing opens the possibility of a variable,
rational, and purposefully oriented action pattern. Gehlen agrees
with Julian Huxley's observation that man's ability to synthesize or
combine every mental trait with any other trait in the area of feeling, knowing, and willing makes a unified mental life possible and
thereby creates culture.

One further consequence of the reduction of instincts in man is
the diffuseness of the original release mechanism of instincts. Konrad Lorenz, who devoted a great deal of scientific effort to the study
of innate release mechanisms in animals, isolated two qualities
which are characteristic components of releasers: maximum sim-
plicity and maximum improbability.[79] Man retained some releasers
insofar as human perception gives preference to those stimuli that

contain the qualities mentioned above. In this sense, human attention becomes structured and oriented. At the same time man is re-
lieved of any natural, uniform, and inevitable response to this stimuli. It is human culture, as Gehlen and Lorenz indicate, which has
taken up the task of providing releasers in the inhibition or facilitation of man's instinctual residues.

What seems to be the case for releasers, seems equally true for
inhibitors. Man, it seems, has lost some of his inhibitions in the
process of becoming human. It is possible that certain inhibitors dis-
appeared because they were compensated for in the form of intelligent action. This might be the case for inhibitors of overindulgence.
The fact that historic man is a killer of his own species may be due
to an instinct residue which became disinhibited. However, intelligent forms of cooperation for much of man's history have been
insufficient to prevent aggression.

In reviewing Gehlen's basic insights and relating them to the
known body of data in social psychology, anthropology, and sociology, it becomes apparent that his theories fit these data better
than any other theoretical argument. They, therefore, hold significant implications for the critical evaluation of sociobiology and
ethology. In contrast to these disciplines, Gehlen has been able to
outline the dynamic and developmental relationships which seem
to exist between human nature and culture. Sociobiologists such as
Wilson, Barash, and van den Berghe have argued that the biological basis of human social behavior is directly related to the motivation principle that individuals will seek to maximize their inclusive
fitness. Some sociobiologists insist that human social organization is
directly predicated on the tendency of members of the human species to increase their reproductive potential within the context of
kinship selection. While these assumptions seem to be incontestable,
when seen in relation to animals, they lack empirical verification
when applied to man.

Ethologists have fared little better in coming to terms with this
problem. The emphasis that such ethologists as Tinbergen, Lorenz,
and Eibl-Eibesfeldt have placed on the existence of fixed action patterns of behavior has provided few insights into man's ability to
develop highly varied and complex forms of behavior which are
culturally specific. In contrast, Gehlen has contributed to the preliminary definition of a multi-dimensional framework of analysis

within which the dynamics that exist between man's biological nature and his capacity for culture may be interpreted from an holistic point of view. It is this framework which may very well provide the groundwork for the development of a new and more comprehensive biosociological paradigm which may be further elaborated and clarified in the future by social scientists, sociobiologists, and ethologists alike.

The Biosociological Theory of Fitness and the Meaningful Unit of Human Evolution

If the logic and empirical validity of our analysis is accepted or at least can be taken as reasonable hypotheses, the following questions may be asked:

1. What is the biosociological definition of fitness?
2. What constitutes the meaningful unit of human evolution?
3. What are the biosociological processes of human evolution?

The term "fitness" as used by ethologists and sociobiologists usually refers to the maximization of the reproductive success of an individual's genes in successive generations. Its opposite meaning, in referring to those individuals who do not reproduce or who do not reproduce at the same rate, is that of being "unfit." Since natural selection is assumed to impinge on the organism of the individual animal, the usage of such a term implies that some organisms are better equipped for survival than others. In the light of what we had to say about human beings, however, the term "fitness" becomes rather ambiguous. The ambiguity stems from the association of "fitness" with existential success and easily jumps from there to the further notion of "right to existence." Although one cannot fault sociobiologists for drawing this conclusion, history has shown that it is easily made.

The term "fitness" as used by us on the following pages neither carries the notion of reproductive success nor refers to the right for existence. It can be argued that the maintenance of a biosocial fitness as we understand it, does not require the maximization of human reproductive success. In fact, extended population growth and the social and environmental pressures that it produces may greatly upset the biosocial fitness of a group. Such behavioral phenomena as infanticide, suicide, homosexuality, celibacy, and organized programs of birth control may be seen as means by which some societies limit their reproductive potential in order to maximize their biosociological fitness. Nor can it be assumed that "fitness" refers to a human individual, in the sense in which it refers to an individual animal. The requirements of "fitness" for animals of the same species living in the environment are the same. If speed is essential for any animal species to survive, females and males, the young and the old must have it, or else perish. In contrast, culture makes it necessary that individuals are biologically different from each other in order that group life be possible. A group of biological geniuses is just as unlikely to produce behavior that makes survival possible as is a group of biological morons. Human culture has its foundation in differentiation and organization which not only allows but requires biological differentiation in order to perform and organize the great number of diverse cultural tasks.

It cannot be assumed that either biological endowment and cultural position for any individual must necessarily be complementary. The effects of the social structure, of traditions, norms, and habits create a great variety of situations where the lack of biological endowment in individuals can be compensated for culturally or vice versa. There is no calculus of natural justice in any culture such that the biologically gifted are also the culturally privileged. Nor need there be such a calculus because cultural creativity by no means is a prerogative of the culturally privileged. On the contrary, cultural creativity proceeds in all areas of cultural life, as it must for a culture to proceed as an integrated whole. Cultural change to a large extent is the emergence of ideas, behaviors, and attitudes that developed somewhere in the social structure, perhaps lay dormant for a long time, and burst into the focus of cultural activities whenever traditional ways and means failed to solve social problems.

The biosocial theorem of fitness therefore can be formulated as follows:

> The fitness of a human group may be interpreted as a condition of dynamic integration among individuals' biological differentiation (in terms of their primary functions), culture, and the ecological

conditions under which they exist, such that behavior is created that makes group survival possible.

The problem of what constitutes the meaningful unit of human evolution has been given extensive attention by a number of social theorists and sociobiologists, inclusive of Marshall D. Sahlins and Elman R. Service in *Evolution and Culture* (1965) and more recently, William H. Durham in his article "The Adaptive Significance of Cultural Behavior" (1976).[80] None of these theorists has pinpointed the evolutionary significance of human culture, nor have they considered the relationship between man's biological nature and the processes of cultural adaptation and change.

From our argument it follows that the meaningful unit of human evolution may be conceptualized as that group of human beings that is *in a position to act on itself in the integration of its genetic potential, its culture, and its environment.*

To some extent, it could be argued that an explanation of the interaction that occurs among man's genetic potential, culture, and the environment, has already been substantially provided by the Pragmatic and Symbolic Interactionist theorists, inclusive of James, Dewey, and Mead. It is of course true that these theorists did provide an overview of how the human self develops within the context of the dynamics of social interaction. Nevertheless, when their assumptions about the biological nature of human nature are closely examined, we find that such assumptions are vague or lack any form of empirical credibility.[81]

For a group to act on itself implies first a degree of sociopolitical control (religious, moral, ideological, political) as well as control over its gene pool. Secondly, it implies a degree of control over the processes of cultural diffusion. Through the exercise of these controls the human group creates the preconditions for a degree of biocultural self-determination which it realizes through the form of integration stated above.

The Biosocial Processes of Human Evolution

One of the major points we have made is that *the selective pressures of the environment are not on the human organism (genes) but on human performance which is the result of the interaction of genes and culture. This performance can be achieved through the integration of varying types of genes and different cultures.* If what in the previous discussion we called the biological nature of the individual is, in fact, the true innate nature of human nature, we might be justified in labelling this package the genetic makeup of the human species. The processes just outlined may then be represented by the following equation:

BIOCULTURAL INTEGRATION TYPE I

$$E - - - - \rightarrow (G \longleftrightarrow C)* = \text{Performance to ensure survival}$$

*E = Environment
G = The genetic makeup
C = Culture
- - - - → = The broken line represents the direction of environmental selective pressures.
←——→ = The continuous line represents the direction of the integration of genetic and cultural factors.

Under these conditions only the *interaction* of G and C would be directly necessitated by the forces of E toward a survival performance. But there would be degrees of freedom allowed for the nature of G and the nature of C. Assuming that E represents some particular ecological niche in which a human group has established itself, the nature of G would have to be such that its interaction with C would make a survival performance possible. However, a whole range of different G's could conceivably interact with a range of different C's to produce such performances. If, for example, the pressures of an environmental situation could be met by close cooperation among the members of a group, the selection toward group oriented genes (expressed in combinations of primary functions) G, as well as the selection for collective cultural practices C in various combinations, would facilitate the survival of the group. The same environmental situation could perceivably be met by different cultural practices integrated with different genetic predispositions.

This integration of G and C holds a number of significant consequences for a group. The environmental responsiveness of the group ensures that the environment will exert selective pressures on whatever forms of cultural and genetic elements the group is in the process of integrating. If the environment happens to be highly specific, as is the case in certain isolated cultures, the integration of genetic predispositions and certain cultural practices would have to be such as to produce performances that would respond to the specific environmental pressures. Therefore, the nature of these cultural practices would tend toward specificity. The selective pressures make it probable that the members of the group will eventually share a narrow and relatively homogeneous genetic range, as they share a relatively uniform culture. The environmental pressures will influence the group's cultural development, such that those practices that have significance for survival are most likely to develop and to be refined. These practices, insofar as they respond to a specific ecological niche and become integrated with a specific genetic composition, ensure that the culture as a whole will develop in a direction that is highly unique.

The selective effects of such a process of biocultural integration may be summarized in the following equation:

EFFECTS OF BIOCULTURAL INTEGRATION TYPE I

$$E_s ----\rightarrow (G_h \longleftrightarrow C_u)^* = \text{Survival performance}$$

$* E_s = $ Environment (specific ecological niche)
$G_h = $ Homogeneous gene composition
$C_u = $ Cultural uniqueness
$----\rightarrow = $ The broken line represents the direction of environmental selective pressures.
$\longleftrightarrow = $ The continuous line represents the direction of the integration of genetic and cultural factors.

The simultaneous integration of a narrow genetic range with a unique cultural configuration, in response to a specific ecological niche has often been observed by anthropologists and forms the basis of all *Gemeinschaft* theories. This cultural condition also contains the basic elements and possibilities for cultural growth. The response determined by the narrow genetic composition and the uniform cultural practices of the group are internal insofar as these genetic and cultural factors adapt to the constraints and demands of the environment. The result is the construction of a socio-cultural organism whose cultural and genetic elements are dynamically equilibrated. At the same time, this response is also active, in that the group in question obtains a degree of control over the environment, however small this control might be initially. In this sense, there exists a feedback relationship between human performance and the environment which can be presented in the following equation:

BIOCULTURAL ENVIRONMENTAL FEEDBACK

$$E_a \cdots\cdots\cdots \leftarrow \cdots\cdots\cdots \leftarrow \cdots\cdots\cdots \leftarrow \cdots\cdots$$
$$\text{Start} \longrightarrow E_s \dashrightarrow (G_h \longleftrightarrow C_u)* = \text{Surplus}$$
$$\text{performance}$$

$*E_s$ = Environment (specific ecological niche)
G_h = Homogeneous gene composition
C_u = Cultural uniqueness
E_a = Altered environment due to human control
$\dashrightarrow$ = The broken line represents the direction of environmental selective pressures.
$\longleftrightarrow$ = The continuous line represents the direction of the integration of genetic and cultural factors.
$\cdots\cdots$ The dotted line represents feedback.

As man's ability to control his environment increases (due to certain crucial inventions which were produced under favorable environmental, genetic, and cultural conditions), the relationship between the environment on the one hand and the genetic-cultural integration on the other hand in part, reverses itself. Although it is unlikely that the human individual will ever have total control over all the environment, for conceptual purposes it might be useful to visualize such a situation through the following equation:

BIOCULTURAL INTEGRATION TYPE II

$$[G \dashleftarrow\dashrightarrow C] \longleftrightarrow E = \text{Surplus performance}$$

Under these conditions, the interaction of G and C would largely determine the nature and the selective effects of E. Control over the environment would lead to a gradual lessening of the natural environmental effects on man's cultural and genetic integration. As the natural environment is replaced by an artificial, man-made environment, the latter already contains so many culture adaptive features that its forces of natural selection are greatly altered. It is of course impossible to obtain a total congruency of man's biocultural integration even within a largely artificial environment. This is due to the fact that the environment changes in response to human control over it and these changes are often unpredictable to the culture. As the environment is never able to totally determine the interaction of human culture and genetic makeup, so human action can never totally determine the environment. In this context the term "biocultural" selection refers to those processes of selection that operate on an environment where human-made conditions predominate which in turn exert selective pressures.

This second type of biocultural integration again holds a number of consequences for the culture-genetic integration of man. A human group that has obtained a degree of control over the environment has greatly altered the selective pressures of the natural environment. The alteration of such selective pressures lessens the dependency of the group on the environment but enhances its dependency on its own cultural creations. Being relatively autonomous from selective pressures of the environment facilitates cultural and genetic diversity within a human group. These processes may be summarized in the following equation:

EFFECTS OF BIOCULTURAL INTEGRATION TYPE II

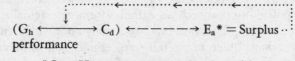

$$(G_h \longleftrightarrow C_d) \longleftarrow - - - - \longrightarrow E_a{}^* = \text{Surplus}$$
performance

$\quad{}^*G_h =$ Heterogeneous genetic composition
$\quad C_d =$ Diversified cultural developments
$\quad E_a =$ Artificial environment (within the constraints mentioned above)
$\longleftarrow - - - - \longrightarrow =$ The broken line represents the direction of the environmental selective pressures.

←——————→ = The continuous line represents the direction of the integration of genetic and cultural factors.
⋯⋯⋯⋯⋯ The dotted line represents feedback.

A surplus performance that results in a group gaining a degree of control over the natural environment enlarges at the same time the cultural scope of the group. With greater cultural diversity there will be selective pressure toward greater genetic diversity. In human populations there are always a great many unusual gene combinations already present in low frequencies. When a performance is required for which a particular gene combination is of great advantage, this can usually be produced through a biocultural integration of gene combinations already present in low frequencies but previously unused by cultural performances. In the second instance, advantageous gene combinations can be obtained through the enlargement of the gene pool by various means.

These broad biosocial processes of human evolution typified in Type I and Type II do not stand in a relationship to each other whereby a progression from I to II is assumed, as underlies the evolution from simple to complex human groups. These two types of biosocial integration are seen as necessary processes operating at any time within the whole range of human evolution, from the simplest to the most complex human group. The cultural and genetic contraction of a group stands in a dialectical relationship to the cultural and genetic expansion of other cosmopolitan groups. It is the culturally and genetically contracting group (Type I - integration) that is a source of never-ending originality and uniqueness in man's evolution. Some human groups carry their originality forward to such extremes as the Pygmies in Africa. There exists the definite possibility that such an extreme form of originality, maintained for any length of time, may be synonymous with a culturally static situation. Other rural, religious, ethnic, and geographically defined groups engage in less extreme forms of cultural and genetic contraction and focussing. The examples of the Hutterites and the Amish in North America indicate that the development of integrative processes of this kind are also possible in highly industrialized societies.

These genetically and culturally contracting and focussing groups are the seedbed of human originality and uniqueness. As such they are indispensable in providing the Type II groups with the basic genetic and cultural elements for growth. The cultural achievements of the cosmopolitan expansive group are made possible through the integration, reshuffling, and recombining of the unique cultural and genetic elements of the Type I groups. The cosmopolitan group is always in danger of losing form and structure. The progressive loss of human originality and uniqueness within a cosmopolitan group leads to a breakup of this group along its peripheries. Inevitably, the cosmopolitan group disintegrates into smaller, culturally and genetically contracting and focussing groups. Human evolution consists of the dialectical interaction between these two types of integration. Each type of integrative process holds important implications for human evolutionary development. The first type of integration has ensured the cultural and genetic differentiation of the human species, in relation to the formation of specific ethnic groups and races. The second form of integration has facilitated the development of cultural expansion and genetic variability among various human groups.

Without the first type of integrative process, the development of the second type would be impossible. Ultimately, the first type of integration must evolve into the second form. Conversely, the second form of integration must evolve again into the first. The significant culture and human genetic changes which have occurred during the course of human history, can be traced to the dynamic interaction that occurs between these two types of integration. To the extent that human performance changes the environment the individual changes himself or herself through such performance. In this sense the place of the human species in the evolutionary process is unique. The human individual is at the same time, the subject and the object of evolutionary development and change.

Conclusion: The Future Significance of Biosociology

There exists a direct line of continuity between the biosociological theories outlined in this paper and some major sociological and anthropological theories. The two types of biosociological integration, as defined here, are linked to the classical theoretical tradition of typology construction. The construction of these typologies is and remains one of the most popular, widely accepted, and fundamental approaches in analyzing social phenomena.

Examples of this classical theoretical tradition in the social sciences include such well-known typologies as Maine's status and contract society; Spencer's militant and industrial forms; Ratzenhofer's conquest state and culture state; Wundt's natural and cultural polarity; Durkheim's mechanical and organic solidarity; Cooley's primary and secondary groups; MacIver's communal and associational relations; Zimmerman's localistic and cosmopolitan communities; Odum's folk-state pair; Redfield's folk-urban continuum; Sorokin's familistic *vs.* contractual relations; Becker's sacred and secular societies and Tonnies's *Gemeinschaft* and *Gesellschaft.* All of these social theorists have indicated that human societies may be ranked according to the level of sociocultural integration which they have achieved.[82] More specifically, it is quite clear that the Type I and Type II forms of integration as defined here, correspond to some degree to what these various theorists have described as either an highly integrated social situation, or a loosely integrated one.

While these typologies constitute major theoretical contributions to the social sciences, they are nevertheless characterized by a number of shortcomings. The most notable of these shortcomings is that the majority of these typologies do not outline in any specific detail the various processes and factors which play a role in the transition from one type of social organization to another. The typologies of social integration, with some exceptions, were formulated as polar opposites. Tonnies's *Gemeinschaft* and *Gesellschaft* is a case in point. No attempt has been made to outline the genetic and environmental dependencies of these types. Nor has any social theorist made the systematic effort to relate these typologies to the processes of cultural change, development, and evolution.

It is only in recognizing the major shortcomings of the classical typological approach to the various forms of social integration that we may come to better appreciate the theoretical significance and usefulness of the biosociological theory of integration. The two types of biosociological integration, outlined in this analysis, are not only precise definitions of the varying types of human social integration but also place the processes of human social integration within an evolutionary biosociological frame of reference. In addition, this biosociological typology directly relates the processes of social integration to the dynamic and multi-dimensional nature of man. Equally as important, this biosociological typology provides a framework for empirical and theoretical research, which may be utilized by evolutionary biologists, biosociologists, sociologists, and anthropologists alike. In short, the biosocial framework outlined here may provide a practical means for the integration of those disciplines concerned with the biological, evolutionary, and socio-cultural aspects of the human species.

NOTES

1. Dennis Forcese and Stephen Richer. "A Sociological Perspective." *In,* Dennis Forcese and Stephen Richer (eds.), *Issues in Canadian Society.* Scarborough, Ontario: Prentice-Hall, 1975, p. 7.
2. Alex Inkeles, *What Is Sociology?* Englewood Cliffs, N.J.: Prentice-Hall, 1964, pp. 14–15.
3. Leonard Broom and Philip Selznick, *Sociology.* New York: Harper & Row, 1963, p. 15.
4. Robert Nisbet, *The Social Bond.* New York: Alfred A. Knopf, 1970, p. 46.
5. Talcott Parsons, *The Social System.* New York: Free Press, 1951, p. 488.
6. See:
    i) Edward O. Wilson, *Sociobiology—The New Synthesis.* Cambridge: Harvard University Press, 1975.
    ii) David Barash, *Sociobiology and Behavior.* New York: Elsevier, 1977.
    iii) Pierre L. van den Berghe, *Man in Society.* New York: Elsevier, 1975.
7. See:
    i) Niko Tinbergen, *The Study of Instinct.* New York: Oxford University Press, 1951.
    ii) Konrad Lorenz, *On Aggression.* New York: Harcourt, Brace & World, 1960.

78   PETER AND PETRYSZAK

iii) Irenäus Eibl-Eibesfeldt, *Love and Hate—The Natural History of Behavior Patterns.* New York: Holt, Rinehart & Winston, 1972.
8. Eliot D. Chapple, *Cultural and Biological Man—Exploration in Behavioral Anthropology.* New York: Holt, Rinehart & Winston, 1970, pp. 4–5.
9. For criticisms of sociobiology and ethology, see:
   i) Marshall Sahlins, cited in "Sociobiology: A New Theory of Behavior" (cover story) in *Time* (Aug. 1, 1977), p. 54.
   ii) Jerome Schneewind, cited in *Time, ibid.,* p. 54.
   iii) John Lev, "A Letter from the Publisher" in *Time, ibid.,* p. 2.
   iv) Sociobiology Study Group of Science for the People, "Sociobiology—Another Biological Determinism," *BioScience,* Vol. 26 (March 1976), p. 182.
   v) American Survey, "Anger Among the Anthropologists," *The Economist* (January 1, 1972), p. 44.
   vi) Ashley Montagu, "On Instincts Again," *Current Anthropology,* Vol. 17 (June 1976), p. 346.
   vii) John A. Searle, "Sociobiology and the Explanation of Behavior." *In,* S. M. Gregory *et al.* (eds.), *Sociobiology and Human Nature.* San Francisco: Jossey-Bass, 1978, pp. 164–82.
   viii) George W. Barlow, "Review of Wilson's *Sociobiology,*" *Animal Behavior,* Vol. 24 (1976), p. 701.
   ix) Roger R. Larsen, "Review Symposium," *Contemporary Sociology—A Journal of Reviews,* Vol. 5 (Jan. 1976), p. 5.
   x) Nicholas Wade, "Sociobiology: Troubled Birth for New Discipline," *Science,* Vol. 191 (March 1976), pp. 1153–54.
10. Barash, *Sociobiology and Behavior,* p. 33.
11. Wilson, *Sociobiology,* p. 3.
12. *Ibid.,* p. 4.
13. See: R. O. Alexander, "The Search for a General Theory of Behavior," *Behavioral Science,* Vol. 20 (1975), pp. 77–100.
14. Wilson, *Sociobiology,* pp. 119–20.
15. *Ibid.,* p. 560.
16. *Ibid.,* p. 560.
17. *Ibid.,* p. 560.
18. Edward O. Wilson, *On Human Nature.* Cambridge: Harvard University Press, 1978, p. 80.
19. Barash, *Sociobiology and Behavior,* p. 81.
20. *Ibid.,* p. 88.
21. *Ibid.,* p. 277.

22. *Ibid.*, p. 318.
23. *Ibid.*, p. 319.
24. *Ibid.*, p. 324.
25. Pierre L. van den Berghe, *Man in Society*, p. 159.
26. *Ibid.*, pp. 63, 203, 233.
27. *Ibid.*, p. 28.
28. See: Iranäus Eibl-Eibesfeldt, *Ethology—The Biology of Behavior.* New York: Holt, Rinehart & Winston, 1975.
29. Tinbergen, *The Study of Instinct*, p. 205.
30. *Ibid.*, pp. 208–9.
31. *Ibid.*, p. 209.
32. N. Tinbergen, "On War and Peace in Animals and Man." *In*, Heinz Friedrich (ed.), *Man and Animal—Studies in Behavior.* New York: St. Martin's Press, 1973, p. 127.
33. *Ibid.*, p. 131.
34. Konrad Lorenz, *Studies in Animal and Human Behavior*, Vol. II. London: Methuen, 1971, p. 193.
35. *Ibid.*, p. 193.
36. Lorenz, *On Aggression*, p. 239.
37. *Ibid.*, p. 246.
38. *Ibid.*, p. 253.
39. *Ibid.*, p. 257.
40. *Ibid.*, p. 265.
41. Konrad Lorenz, *Evolution and Modification of Behavior.* Chicago: University of Chicago Press, 1965.
42. *Ibid.*, pp. 107–8.
43. Eibl-Eibesfeldt, *Love and Hate—The Natural History of Behavior Patterns*, p. 3.
44. *Ibid.*, p. 4.
45. *Ibid.*, p. 5.
46. *Ibid.*, p. 6.
47. *Ibid.*, p. 32.
48. *Ibid.*, p. 32.
49. *Ibid.*, p. 100.
50. W. F. Bates, "The Disturbing Political Implications of Sociobiology," *Science Forum*, Vol. 9 (Oct. 1976), p. 24.
51. Peter K. Smith, "Biosocial Anthropology," *British Journal of Social and Clinical Psychology*, Vol. 15 (1976), p. 450.
52. Arthur Caplan, "Sociobiology," *Philosophy of Science*, Vol. 43 (1976), pp. 305–6.
53. Fred Templeton, "Man in Society," *American Journal of Sociology*, Vol. 82 (Nov. 1976), p. 708.

54. Edward Tiryakian, "Biosocial Man—Sic et Non," *American Journal of Sociology, ibid.,* p. 705.
55. Joseph S. Apler, "Biological Determinism," *Telos,* Vol. 5 (Spring 1977), p. 172.
56. David W. Paulsen, "Sociobiology—Review," *Theory and Decision,* Vol. 7 (July 1976), p. 240.
57. See:
    i) Eileen Barker, "Sociobiology," *British Journal of Sociology,* Vol. 24 (1975), p. 501.
    ii) Bruce K. Eckland, "Darwin Rides Again," *American Journal of Sociology,* Vol. 82 (Nov. 1976), pp. 692–97.
58. Marshall Sahlins, *The Use and Abuse of Biology.* London: Tavistock, 1977.
59. *Ibid.,* p. 57.
60. *Ibid.,* p. 57.
61. Robin Fox and Usher Fleising, "Human Ethology," *Annual Review of Anthropology 1976,* Vol. 5 (1976), p. 265.
62. See:
    i) A. Roe and G. C. Simpson (eds.), *Behavior and Evolution.* New Haven: Yale University Press, 1958.
    ii) T. Dobzhansky, "Cultural Direction of Human Evolution," *Human Biology,* Vol. 35 (1963), pp. 311–16.
    iii) M. F. A. Montagu (ed.), *Culture: Man's Adaptive Dimension.* New York: Oxford University Press, 1968.
63. Robert Bigelow, *The Dawn Warriors—Man's Evolution Toward Peace.* Boston: Little, Brown, 1969, p. 243.
64. William H. Durham, "The Adaptive Significance of Cultural Behavior," *Human Ecology,* Vol. 4 (1976), p. 90.
65. This overview of Gehlen's theories is based on his major work: *Der Mensch—Seine Natur and seine Stellung in der Welt.* Wiesbaden, Germany: Athenaion, 1976.
66. See: A. Gehlen, *Studien zur Anthropologie und Soziologie.* Berlin: Tuchterland, 1963; *Moral und Hypermoral.* Wiesbaden: Athenaion, 1973; *Urmensch und Spatkultur.* Wiesbaden: Athenaion, 1975; *Der Mensch.* Wiesbaden: Athenaion, 1975.
67. Lorenz, *Studies in Animal and Human Behavior,* Vol. II, p. 192.
68. L. Bertalanffy, *General Systems Theory.* New York: Braziller, 1969.
69. P. Alsberg, *In Quest of Man.* New York: Pergamon, 1970.
70. See: Bernard G. Campbell, *Human Evolution—An Introduction to Man's Adaptations.* Chicago: Aldine, 1966, pp. 48–49.
71. See:
    i) G. Allport, *Personality—A Psychological Interpretation.* New York: Holt, Rinehart & Winston, 1973.

    ii) G. Lindzey, and C. S. Hall (eds.), *Theories of Personality.* New York: John Wiley, 1965.

72. Erik Erikson, *Identity: Youth and Crisis.* New York: Norton, 1968, p. 94.
73. *Ibid.,* p. 94.
74. *Ibid.,* p. 115.
75. See:
    i) E. Faris, "The Primary Group: Essence and Accident," *American Journal of Sociology,* Vol. 38 (1932), pp. 41–50.
    ii) G. C. Homans, *The Human Group.* New York: Harcourt, Brace, 1950.
    iii) E. A. Shils, "Primary Groups in the American Army," *In,* R. K. Merton and P. L. Lazarsfeld (eds.), *Continuities in Social Research.* New York: Free Press, 1950, pp. 16–39.
    iv) A. Keith, *A New Theory of Human Evolution.* Gloucester, Mass.: P. Smith, 1968.
    v) T. M. Mills, *The Sociology of Small Groups.* Englewood Cliffs, N.J.: Prentice-Hall, 1967.
76. M. Landmann, *Philosophical Anthropology.* Philadelphia: Westminster Press, 1974.
77. G. Pfahler, *Vererbung als Schicksal.* Leipzig: Johann Ambrosius Barth, 1932.
78. H. Hoffman, *Das Problem des Charakteraufbaus.* Hamburg: Verlag van Julien Springer, 1926.
79. Lorenz, *Studies in Animal and Human Behavior,* Vol. II, p. 193.
80. Marshall D. Sahlins and Elman R. Service (eds.), *Evolution and Culture.* Ann Arbor: University of Michigan Press, 1965, p. 53.
81. Nicholas G. Petryszak, "Biosociology of the Social Self," *Sociological Quarterly* 20 (Spring 1979): 291–303.
82. J. C. McKenney and C. P. Loomis, "The Typological Tradition." *In,* F. Tonnies, *Community and Society.* New York: Harper and Row, 1965.

THE GENETICS OF SOCIOBIOLOGY

The Origins of Sociobiology

In the substantial volume that Edward O. Wilson published in 1975 he presented sociobiology as a comprehensive theory of evolutionary biology which would eventually "cannibalize" ethology and comparative psychology and dominate biology above the molecular level. This "New Synthesis" was achieved by putting together several concepts from earlier streams of thought. One of these was the notion of a genetically determined society as exemplified by the social insects. This subject had been Wilson's specialty and, to some extent, sociobiology resulted from extending entomological sociology to embrace the vertebrates. A second thread woven into the new fabric was the idea of kin selection. From the time of Darwin's *Origin* there had been considerable puzzlement over how to explain the development of self-sacrifice by natural selection. For social organization to function, the individual must sometimes curb his own drives in deference to others, on occasion, even to the point of giving his life. How can giving one's life increase the probability of leaving descendants? Darwin recognized this paradox in the case of the social insects (1859, pp. 237-38) and decided it could be explained by selection favoring members of the family. The sterile worker bee, by devoting her life to the efficient functioning of the hive, would raise the probability that non-sterile sister queens, with heredity very close to hers, would found other hives. In post-Darwinian times this problem was generally glossed over by picturing selection as favoring "the good of the species."

In general, this was the position of the ethologists—the students of animal behavior typified by Lorenz—but they accepted a third idea also accepted by Wilson: that animal behavior, because it had been formed by natural selection, was adaptive and, therefore, genetically based. Species-specific behavior was the result of innate releasing mechanisms.

In 1962 V. C. Wynne-Edwards in his *Animal Dispersion in Relation to Social Behavior* carried the "good of the species" idea so far that a strong reaction arose against it. He argued that individual animals reacted to increasing population density by decreasing their reproductive activity and thus were able to maintain a balance between population and resources. His critics—for example, G. C. Williams in *Adaptation and Natural Selection* (1966)—made clear that for selection to change heredity within the species, it must act between competing individuals and not between deferential groups. Thus the old question of how to account for the selection of hereditary altruism came up again.

In 1955 J. B. S. Haldane had considered this problem in a somewhat relaxed mood. "Let us suppose that you carry a rare gene," he wrote, "which affects your behavior so that you jump into a flooded river and save a child, but you have one chance in ten of being drowned, while I do not possess the gene and stand on the bank and watch the child drown. If the child is your own child or your brother or sister, there is an even chance that the child will also have this gene, so five such genes will be saved in children for one lost in an adult. If you save a grandchild or nephew the advantage is only two and a half to one. If you only save a first cousin, the effect is very slight. If you try to save your first cousin once removed the population is more likely to lose this valuable gene than to gain it. . . . It is clear that genes making for conduct of this kind would only have a chance of spreading in rather small populations where most of the children were fairly near relatives of the man who risked his life. It is not easy to see how, except in small populations, such genes could have been established. Of course the conditions are even better in a community such as a beehive or ants' nest, whose members are all literally brothers and sisters" (*New Biology* 18:34).

What Haldane described is the idea of kin selection. But it remained for W. D. Hamilton, in two papers in the *Journal of Theoretical Biology* in 1964, to establish and give great prestige to the principle of kin selection by providing it with a more elaborate mathematical formulation and a supplemental concept—inclusive fitness—the direct contribution of one's own genes to the next generation plus the increased probability of their survival as a result of one's own behavior either before or after the production of offspring.

Hamilton began his argument with the ants and the bees and emphasized that worker ants—or bees—not only were of the same colony sisters, but, because of the peculiar sex-determining mechanism in hymenoptera, had three-quarters of their genes in common instead of one-half, the figure for sisters in almost all other animals and the figure that would apply to their own daughters if they were to have any. Hence they increased their inclusive fitness by caring for their sisters rather than having offspring of their own who would be more distantly related to them. It was for this reason, Hamilton argued, that social behavior had developed in the hymenoptera on several occasions independently. Social behavior had developed among birds and mammals in the absence of the facilitation provided by the peculiar hymenopteran chromosomal system because, as Haldane had pointed out, it was favored by the situation existing in small groups where the proportion of close relatives was high. Hamilton applied the same general mathematical formulation to both cases.

The concept of inclusive fitness was further developed by R. L. Trivers. First (1971) he formulated the idea of selection for reciprocal altruism—aid rendered to a non-relative when the chances of reciprocation are good enough to enhance one's inclusive fitness. Then he took up parent-offspring relations, raising the question (1972) of what degree of parental investment in a given offspring would maximize the inclusive fitness of the parent. Later (1974) he considered the conflict inherent in the attempts of both offspring and parent each to maximize its own inclusive fitness.

Wilson's 1975 volume constitutes an attempt to synthesize all these elements: the biological basis of social behavior, its origin in phylogeny, its production by natural selection, and the presence of a genetic basis explained by kin selection and the individual's striving to maximize its inclusive fitness.

The explosion of criticism which greeted the publication of the book (Sociobiology Study Group and Wilson, 1976, *Bioscience*) and the polemic which has followed (Harris & Wilson, 1978; Wade, 1976) have in general centered on the emphasis that the theory places on the genetic component in behavior. Since 1975 there have been further contributions from the sociobiologists. Wilson's *On Human Nature* (1978) is definitely less abrasive than *Sociobiology* and appears designed to dissociate the author from the most extreme genetic determinism. Other authors, J. Maynard Smith and G. A. Parker, for example, have invoked game theory to explore the problem of how selection deals with cheaters in reciprocal altruism. R. Dawkins has apotheosized inclusive fitness by applying it to the single gene, *The Selfish Gene,* (1976). "The genes march on . . . That is their business," he says. "They are the replicators and we are their survival machines. When we have served our purpose, we are cast aside. But genes are denizens of geological time: genes are forever" (p. 37). "As far as a gene is concerned, its alleles are its deadly rivals" (p. 40). "Like successful Chicago gangsters, our genes have survived, in some cases for millions of years, in a highly competitive world. This entitles us to expect certain qualities in our genes. I shall argue that a predominant quality to be expected in a successful gene is ruthless selfishness" (p. 2).

Not all sociobiologists agree on everything. Dawkins would excommunicate anyone caught considering the possibility of group selection. Wilson thinks it may operate on occasion. But by and large there seems to be a strong consensus, tending to solidarity under external attack, and the unifying basis of the common *Weltanschauung* is a belief that cooperative behavior is strongly influenced by genetic factors. It seems worthwhile, therefore, to ask whether the genetic concepts accepted and the evidence cited by the sociobiologists are consistent with current developments in population and molecular genetics.

Single Gene vs. Polygenic Inheritance

In all the writings of the sociobiologists the fundamental genetic concept is the single Mendelian gene influencing behavior. In chapter 5 of *Sociobiology* Wilson speaks again and again of the increase in frequency of an altruistic gene in a population, of the possibility of its becoming fixed—reaching 100%—of whether it is dominant or recessive. In chapter 27, "Man: From Sociobiology to Sociology," he writes, (p. 554) ". . . if a single gene appears that is responsible for success and upward shift in status, it can be rapidly concentrated in the uppermost socioeconomic classes." In *On Human Nature* (p. 47) Wilson rejects the notion of "mutations for a particular sexual practice or form of dress," and agrees that the "behavioral genes more probably influence the ranges of the form and intensity of emotional responses," but he believes "that the position of genes having indirect effects on the most complex forms of human behavior will soon be mapped on the human chromosomes."

Similar thinking in terms of the selection of single genes appears to be characteristic of most of the members of the sociobiology group, not merely Dawkins, Hamilton, and Trivers but also of the proponents of game theory, such as G. A. Parker (1978) and John Maynard Smith (1978) who contend that selection produces adaptive evolutionarily stable strategies (ESS) for handling conflicts within the population. "The mixed strategy that is evolutionarily stable for the Hawk-Dove game," writes Maynard Smith (p. 186), "is to play hawk with probability 8/13 and play dove with probability 5/13. I shall not discuss the derivation of this strategy here, but it is not difficult to see that the strategy does fulfill the second requirement of being evolutionarily stable against, say, a mutant hawk strategy."

Picturing genes as hard, discrete units, varying only in quantum mutational jumps, as the sociobiologists do, has its origin, of course, in the Mendelian tradition, particularly as interpreted by H. J. Muller, and is presently characteristic of the school of mathematical geneticists of which Crow and Kimura (1970) are representative. Those who think in these terms generally assume that there is one best genotype composed entirely of normal alleles and that variants are either defects produced by mutation and in the process of being eliminated by selection, or are selectively neutral alternate forms. Heterozygosity is, therefore, temporary or transitional. Any advantageous mutant appearing in such a population will replace its less favored predecessor. The case of a heterozygous combination of two alleles being superior to either homozygote is so rare that it plays no important role in evolution. It is this school of thought that has developed the concept of the genetic load, the degree to which the fitness of a population is depressed below what it would be if it contained no unfavorable alleles, if every individual had the best possible genotype.

There is, however, another way of viewing genetic phenomena which many geneticists believe is closer to reality and which stems from the work of Dobzhansky (1970), Lerner (1954), and Mayr (1963) and is presently represented by Lewontin (1974) and Wallace (1968). It emphasizes the interaction between genes, doubts the possibility of assaying the contribution of every gene to the individual phenotype, and sees selection as acting on groups of coadapted alleles rather than on exhibitionistic, selfish genes.

The protagonists of these two points of view might be called the uniformists and the variabilists. The former, with whom the sociobiologists identify themselves, think of the gene pool—the total number of alleles in the present population that are thus available for the formation of succeeding generations—as constantly being purged by selection of the defective alleles produced by mutation and thus approaching uniformity with all alleles of the best possible type. The variabilists, on the other hand, reject the idea of one best genotype. In any randomly breeding population, they contend, fitness is correlated with the modal phenotype—those individuals with characters approximating the mean. The extreme phenotypes for whatever characters—height, weight, proportions—taken together, contribute fewer offspring to the next generation than those in the middle of the distribution. But the individuals near the mean, who are phenotypically alike, are by no means genetically uniform. A comparison of the offspring of two mated pairs from this group will show not merely differences between parent and offspring but different degrees of variation within the offspring of different parents.

To the variabilist what selection appears to be doing, instead of picking out genes for their individual excellence, is preserving a variety of gene combinations in individual chromosomes such that when these chromosomes are randomly combined to produce new individuals, the new combinations will have a high probability of producing individuals of the modal phenotype. In every generation random recombination will produce some extreme phenotypes, but these will be rare. In fact, where it has been looked for, evidence has been found that extreme phenotypes are less common than would be expected by chance. In a population at equilibrium the distributions of characters are what is termed leptokurtic, for a higher proportion of individuals is clustered near the mean than would be found in a normal distribution.

Deviation from the modal phenotype is also commonly associated with decreased fitness—viability or fecundity or both are reduced—and these phenotypic non-conformists contribute fewer offspring per individual to the next generation than their more average contemporaries. This process keeps the modal phenotype of the population stable. If artificial selection is practiced, one can drive the mean of a character above or below the population mean by limiting the parents to the upper or the lower part of the distribution. But if this process is continued, the amount of the response will become less as the number of generations increases and the general fecundity of the population will decline. If selection is then discontinued, and the population is allowed to breed at random, the means of the characters changed by selection will tend to return to their former values and at the same time the fecundity will recover.

Observations of this sort are familiar to animal breeders and have been described by Lerner (1954) as illustrating the principle of genetic homeostasis. The gene pool of a population is not merely an aggregate of individual genes. It is a co-adapted system of allelic frequencies which, under the breeding system followed, will keep the modal phenotype within the same limits generation after generation. At most loci there is more than one allele. The complete replacement in a population of one allele by another is a rare occurrence and is probably not the way evolution proceeds. Where it has been possible to analyze the chromosomes of selected animals for the location of the alleles responsible for a change in the mean of a measurable character, the results have indicated a large number of different sites distributed throughout the chromosome complement.

On pages 70 to 73 of *Sociobiology* there is a brief discussion of the maintenance of genetic variability in populations, and there are citations to Milkman, Lewontin, Lerner, and Waddington. But in fact, one could steep oneself in the literature of sociobiology and never suspect that the variabilists represent a very important stream of genetic thinking. This is particularly unfortunate because very recent developments in molecular genetics promise to give the variabilists new and powerful evidence for their position and to deepen and broaden our understanding of the whole field of genetics in a truly revolutionary way.

Recent Developments in Molecular Genetics

During the past twenty years the uniformists and the variabilists have been engaged in a frustrating polemic over how much genetic variability there is and how to measure it. Most genes direct the assembly of chains of the twenty amino acids in precise sequences. These chains—varying in length from a few dozen to some hundreds of amino acids—are termed polypeptides and, either alone or in combinations, they constitute proteins, the most important molecules in determining the structure and function of an organism.

Hemoglobin is a protein composed of four polypeptide chains, two of one kind, alpha, 141 amino acids long, and two of another kind, beta, of 146 amino acids. If the sixth amino acid in the sequence of the beta chain is changed from glutamic acid, which is normal, to valine, the individual having beta chains of this altered type only (termed hemoglobin s) has sickle cell disease and is very ill. In the 1960s it was discovered just how, in bacteria, a gene specified a given amino acid at a given point in the peptide sequence. It is done by a sequence in the DNA of three of the four possible nucleotides: A, T, G, C. These are transcribed into the complementary nucleotides, U, A, C, G of RNA in a molecule called messenger RNA (m-RNA) which then directs the appropriate amino acid into the peptide chain. This genetic code was found to apply in higher organisms as well. If the DNA reads CTT in the sixth position of the gene for the beta chain, the transcribed RNA will read GAA and a glutamic acid will go into the sixth position. If the second nucleotide of the DNA is changed to A, the corresponding RNA triplet will read GUA and will direct valine into the sixth position. Hemoglobin s can be recognized because it produces severe symptoms in those who have no other type of beta chain and it can also be identified in the laboratory because it moves to an identifiable position when subjected to an electric field in a process known as electrophoresis.

When one attempts to identify allelic variants, one can begin by listing those, such as hemoglobin s, which show marked functional defects. Most of these can also be identified by electrophoresis. But there are many cases where altered electrophoretic behavior is not accompanied by obviously altered function; these are added to the functional variants in making the census of variability. This has proved to be not an entirely satisfactory procedure. Electrophoretic variants without other phenotypic effect may not come to light. Furthermore, some amino acid substitutions do not alter the electrophoretic properties of the molecule nor do they produce other obvious changes in phenotype. Many of these are undoubtedly overlooked. The question whether alleles that have no demonstrable effect on the phenotype are or are not selectively neutral is very nearly impossible to resolve. To compound the confusion, it has been shown that for some proteins, the conditions under which electrophoresis is carried out can change the classification of a given molecule from normal to variant (Singh *et al.*, 1976).

But to define an allele as a sequence or DNA producing an observably different protein product, which electrophoretic identification requires, is to ignore much that is well known about biological processes, especially development and metabolism. Biological molecules interact, not chaotically, but according to a program which guides development and maintains metabolism within the limits that characterize the healthy state. This is achieved by a complicated system of feedbacks which render the organism a self-sustaining, self-reproducing cybernetic device. To be a part of such a system a gene must do more than produce a product of high specificity. This product must appear at the proper time in development, in the proper amount, in only the proper cells of the organism and only when the metabolic state of the organism calls for its production.

Long before the precise nature of gene action was known, it was recognized that a given locus could have alleles that were clearly mutant as well as others less different from the wild-type, some of which approached the wild-type very closely. Alleles at the white locus in Drosophila give eyes varying from pure white to close to the normal brick red. The C locus in the mouse can produce albinos or individuals only slightly paler than mousy. As early as 1943 Stern and Schaeffer published evidence that even normal alleles could be shown to be subtly different from each other. These they termed iso-alleles.

The understanding of the transfer of genetic information from DNA to m-RNA by transcription and from m-RNA into protein by translation was achieved in the 1960s through work with bacteria. In these simple prokaryotic organisms DNA is not segregated in a nucleus. In general, gene action is controlled either by repressor proteins which prevent transcription unless they are inactivated by inducers, or by proteins which allow transcription unless they are converted into repressors by specific molecules in the environment. Translation begins at one end of the m-RNA strand before transcription has completed the other. Much effort has been expended, but without success, in trying to find the bacterial types of controls in the higher eukaryotic organisms where the DNA is confined in a nucleus, where internal cellular control must be integrated with intercellular reactions and where messages must be transmitted across membranes—nuclear within the cell and plasma between cells.

In the eukaryotes one finds m-RNA in the cell outside the nucleus, and it is translated into protein there by a process very similar to that observed in bacteria. But there is no m-RNA in the nucleus; instead, there is a snarl of RNAs of enormously varying length. In bacteria, there are no indications of repetitions in the sequences in the DNA. In eukaryotes, on the other hand, forty-some per cent of the DNA consists of sequences repeated from many to hundreds of thousands of times and these repeated sequences never seem to be transcribed into m-RNA. These facts have been known for more than a decade and have constituted a frustrating puzzle.

The foregoing discussion of gene action and cellular controls

covers work done largely in the 1950s and 1960s by so many in-
vestigators that to cite the sources would obscure the narrative. For
a review of this earlier work see Stent (1978) or Judson (1979).
Very recent experimental data promise a way of interpreting the
formerly puzzling facts and of fitting them into a comprehensive
theory of genetic control in higher organisms. It is worthwhile going into this very recent work in some detail. In 1977 two papers
appeared reporting startling observations on eukaryotic RNA and
DNA. In July, Bastos and Aviv published in the journal *Cell* data
establishing that in the precursers of mouse red-cells, globin m-
RNA is produced in the nucleus first as a very long strand which
is processed in two stages and then exported to the cytoplasm with
the length which is translated there. Four months later in *Nature,*
Breatnach, Mandel, and Chambon described the DNA coding for
the protein ovalbumin which makes up much of the white of egg.
They found that the sequence coding for ovalbumin is not continuous, but is interrupted by extensive sequences which are not
represented in the m-RNA and hence are never translated into protein. Since then other papers have confirmed these findings and extended them to other genes. Other papers keep pouring from the
laboratories (Crick, 1979). It is now possible to describe the probable structure of eukaryotic genes and their production of m-RNA
in a way that strongly suggests mechanisms of gene regulation.

The DNA of eukaryotes contains a great deal of information in
addition to that which determines the sequences of amino acids in
proteins. Any sequence—gene—coding for a protein is interrupted
one or more times by extensive sequences that are never reflected
in protein composition, and similar sequences flank the gene on
either side. The gene for the beta chain of human hemoglobin,
which needs only 438 nucleotides to code for its 146 amino acids
has an intervening sequence of 1000 nucleotides and is separated
from the adjacent delta gene by 7000 nucleotides (Flavell *et al.,*
1978, p. 25). In this case the intervening and flanking sequences
are about ten times the length of the sequence that is transcribed
into m-RNA.

When a gene is transcribed, the intervening sequences and some
part of the flanking sequences are transcribed with it. This results
in an RNA molecule many times longer than the functional mes-

senger which is produced by trimming and excising parts of the much longer original. The DNA sequences which do not code for amino acids undoubtedly include the repetitive sequences and they add up to a very substantial part of the genome. Such sequences are expensive for the organism to maintain. They have to be copied whenever a cell replicates and this requires substantial quantities of energy and materials. No species or organism could afford to go on carrying such a burden if it did not contribute vitally to individual survival. Primates have lost the sequences coding for the enzyme which produces ascorbic acid. They depend on getting it in their diet. But they hold on to their repetitive sequences.

It is difficult to reject the probability that this apparently super-fluous information carries the messages that determine when, where, under what conditions, and in what amount a gene is to be transcribed and translated. In fact, there is already substantial evidence that this is the case. In addition to the many mutant alleles that alter the amino acid composition of the hemoglobin chains—either alpha or beta—there are others, known as thalassemias, which leave the amino acid sequences completely untouched, but instead, drastically reduce the number of chains produced. Nothing is wrong with the information specifying the amino acids. Something is amiss somewhere else. Where can it be? One very likely place is the untranslated sequences. There are other hemoglobin mutants in which the normal amino acids are present in the proper sequence, but thirty-some additional amino acids are tacked onto the end of the chain. This must result from normally untranslated sequences being translated. Three different mutants of this type are known and in all three cases the abnormally long chain is produced in sharply reduced quantity. Apparently, when some of the information usually untranslated functions abnormally, it can change the amount of protein produced. It will be surprising if, within the next year, someone does not demonstrate that one of the thalassemias results from changes in either the intervening or the flanking sequences of the globin gene involved.

These very recent developments in molecular genetics lend powerful support to the position of the variabilists. Variability, and hence heterozygosity, must be found not only in the sequences of amino acids but also in the differences in the nucleotide sequences of the untranslated portions of the DNA. Since for every gene there are thousands of nucleotides in such sequences, the potential for variation becomes enormous—variation which is never going to show up in electrophoretic differences between proteins. The possibility of dozens if not hundreds of alleles at every locus is at once apparent and the possibility of different combinations of alleles between loci approaches the infinite. By using only a little imagination one begins to comprehend how selection, with such an instrument to play on, can orchestrate a modal phenotype with many genotypes and yet hold it constant, can retain within the population large numbers of different alleles which work effectively when paired in single individuals and can select these effective pairs for the way they cooperate with other effective pairs at other loci. This is precisely the type of genetic system which the variabilists have found themselves forced to postulate from their experimental observations. It leads to conclusions quite different from those of the single-gene uniformists which characterize sociobiology. Instead of discrete, indestructible, agressive units, constantly urging their carriers to hand them on at any cost, one must visualize fragile combinations of continuously varying, interacting complexes under constant threat of vanishing as a result of recombination.

Genetic Determinism

"Among persons in the general population, the concentrations of lipids and lipoproteins in plasma show a continuous distribution. When one defines the upper five per cent of a population as being abnormal for plasma cholesterol, triglycerides or a specific lipoprotein fraction, no more than one out of five such hyperlipidemic subjects has a single-gene determined form of hyperlipoproteinemia. Rather, the majority of hyperlipidemic subjects owe their abnormality to a complex interaction of polygenic and environmental factors." This statement by Fredrickson, Goldstein, and Brown (1978, p. 604), three of the outstanding authorities on abnormal lipidemia, makes clear that in the total population only one person in a hundred is pathologically hyperlipidemic, and thus at high risk for coronary disease, because of single-gene action. Four times as many people are similarly abnormal for other causes. Similar statements could be made about numerous other human afflictions such as gout or duodenal ulcer, and few medical geneticists are any longer under the compulsion of finding single-gene explanations for human ills.

Since single-gene genetics fails to explain a very large part of metabolic pathology, it is very likely that its role in determining behavior is even less substantial. Yet Wilson insists that human rules against incest have a genetic basis. Without doubt he holds this conviction deeply, for he expresses it in both *Sociobiology* and *Of Human Nature,* and he defended it vigorously in his debate with Harris (Harris & Wilson, 1978, p. 13). I cannot find a reference to "the incest gene" but on this matter he says, "human beings are guided by an instinct based on genes" (*Of Human Nature,* p. 38).

The effects of heterozygosity are advantageous, and many plants and animals have developed means of guaranteeing outbreeding to preserve them. But outbreeding is not universal. Close inbreeding in a human population has deleterious consequences. But the development of institutionalized rules for marriage seems much more likely to have arisen from the fact that human kinship formed the basis of social organization and personal status and that to maintain the entire fabric required proscribing certain unions and requiring others. If the basis was biological, it is hard to explain why relatives of the same degree are forbidden to marry in some cultures and encouraged to do so in others. Incest rules have generally favored marriages between different groups and thus increased the solidarity of the larger group by establishing a network of personal relations between smaller groups. In those cases, such as Pharaonic and Ptolmaic Egypt and Incan Peru, where incest was the rule in the royal line, it had the effect of setting royalty off from the rest of the population and emphasizing its very special character.

The strong penchant of the sociobiologists to think in terms of genetic determinism and single-gene effects is accompanied by a reduced interest in the behavioral development of the individual. All adaptive behavior occurs as a reaction to an environment. If the environment is aberrant, the behavior may never be elicited or it may be performed without achieving anything. A lepidopterous larva, programmed to pupate underground, will, after finishing feeding, engage in exploratory crawling over the ground and then burrow into the soil. But such a larva placed in a metal container will crawl mechanically for hours around the edge of the floor until internal changes toward the prepupal state make it impossible for him to crawl further. He will never burrow. Drosophila, ten or twelve hours after eclosion from the pupa will begin to court— engage in a series of signals and responses which lead to copulation. If only one sex is present, signals elicit no appropriate response and the activity leads nowhere.

In higher animals a much more complex nervous system has developed, and the individuals are capable of much more complex behavior. They have a wider repertory and are able to counter blocks and overcome frustrations to a degree not possible for simpler creatures. The central nervous system of higher vertebrates actually functions as a computer. The human brain represents the ultimate development in this direction and is certainly the most versatile computer ever devised or constructed. The advantage of a computer is versatility and plasticity in problem solving. Rigid, ready-made answers do not have to be built into it. On the basis of general rules and stored information gleaned from experience it can generate novel ways of solving problems. Once this type of plasticity reaches a certain level, it becomes more adaptive for selection to increase the plasticity and the versatility and to reduce the number of rigid, ready-made answers built into the program. It seems much more likely that selection has worked toward greater flexibility in human behavior rather than toward amassing a great aggregation of rigid imperative minutiae.

The increase in complexity of the nervous system is reflected in a slower maturation, and experience can actually affect the functioning and the interactions of nerve cells in the brain of the neonate mammal (Barlow, 1975). Similarly, experience and reaction guide the developing animal in the achievement of a personality with a unique internal integration and a similarly unique set of social relations and behavior. This does not mean that the individual genotype is of no consequence in the formation of the total animal, but it does mean that the phenotype, which *is* the total animal, is actualized by the whole history of interactions which have taken place between the genotype and the environment. It is a part of

the folklore that a wild animal artificially raised from birth by man
is not the same as one that grew up in its natural habitat. Konrad
Lorenz (1965, p. 80), who shares with the sociobiologists a belief
in the primacy of genetic influences on behavior, recognized that
the experimenter, when looking for evidence of innate behavior,
must be sure that his animals have not been subjected to "bad
rearing."

Probably the clearest illustration of the complex interaction between genetically determined potential and specific actualization is
provided by human language. The human genome contains information which endows the individual with the ability to learn and
use language. Language consists of the recognition of symbols and
of a system of relations between them based, at least initially, on
vocal elements sensed by the ear. No other animal has this ability
with anything approaching the competence and the virtuosity of the
human being. But genetics has no influence on what language the
individual will speak. By the time a child is five he has learned
the elements of the language spoken around him regardless of what
language his biological parents spoke.

The only child who ever learned a language in isolation from all
human contact was Tarzan of the Apes, and he seems to have been
unique. There is a childish desire to believe, which has been around
for at least 2400 years since Thucydides described it, that if a
group of children could be raised in complete isolation, they would
come to speak to each other in the original, genetically determined
language. Here is the *reductio ad absurdum* of genetic determinism
—the gene as a magic talisman which at the appropriate, predestined moment will cause to bloom a whole complex system of
coordinated behavior.

I hesitate to attribute such a belief to the sociobiologists, but on
the basis of Wilson's writings it is difficult to let them off the hook.
In *Sociobiology* (p. 560) Wilson quotes with approval Robin Fox's
statement that if children raised in isolation, as in the reputed experiments of the Pharaoh Psammeticos and James IV of Scotland,
had survived, "I do not doubt they *could* speak and that, theoretically, given time, they or their offspring would invent and develop
a language despite their never having been taught one." How is
one to react to such a reckless refusal to recognize the power of

delicate interaction, not only between parent and offspring, but also between successive cohorts? This is the equivalent of expecting that the pure DNA extracted from a few human genomes will, if sprinkled on the ground, repeople an uninhabited planet.

This rejection of ontogenetic influences is not confined to Wilson. It pervades the sociobiological literature. Trivers (1971) in "The Evolution of Reciprocal Altruism" says: "No careful work has been done analyzing the influence of environmental factors on the development of altruistic behavior, but some data exist. . . . Such personality differences are probably partly environmental in origin."

In Wilson's discussion of incest, previously mentioned, there is another passage showing how strongly the possibility of environmental influence is rejected. Wilson cites a study by Joseph Shepher (1971) on relations between children raised together in Israeli kibbutzim. Shepher finds no marriages among such children and concludes that an inhibition against sexual involvement arises automatically in children who are intimately associated during preadolescence. This study, as published in *Archives of Sexual Behavior* is rather episodic and not very convincing. Wilson finds it compelling, but for him the experience of association among the young children is of secondary importance. "Bond exclusion of the kind displayed by the Israeli children," he writes (1978, p. 38), "is an example of what biologists call a proximate (near) cause; in this instance, the direct psychological exclusion is the proximate cause of the incest taboo. The ultimate cause suggested by the biological hypothesis is the loss of genetic fitness that results from incest. . . . To put the idea in its starkest form, one that acknowledges but temporarily bypasses the intervening developmental process, human beings are guided by an instinct based on genes." Could the doctrine of genetic determinism be more unequivocally embraced?

Emphasis on Conflict and Violence

Grafted to the single-gene analysis and genetic determinism there is in sociobiology an emphasis on conflict and violence. This can to a considerable extent be traced to Trivers's writings (1972, 1974). In the former he says, "To discuss the problems that confront individuals ostensibly cooperating in a joint parental effort, I choose the language of strategy and decision, as if each individual contemplated in strategic terms the decision it ought to make at each instant in order to maximize its reproductive success. This language is chosen purely for convenience to explore the adaptations one might expect natural selection to favor" (p. 146). This has resulted in the introduction into the literature of an emotionally loaded vocabulary containing such words as "desertion" and "cuckoldry" and a general aura reminiscent of Restoration comedy. A study (Powers, 1975) has stigmatized the bluebird, traditional greeting-card symbol of disinterested solicitude, as inherently selfish because when the experimenter removed one member of a pair with nestlings, the substitute mate who moved in to fill the gap ignored the existing offspring.

Wilson writes (1975, p. 342), "Trivers' hypothesis [of parent off-spring conflict] is consistent with the time course of conflict observed in cats, dogs, sheep and rhesus monkeys. In each of these species the conflict begins well before the onset of weaning and tends to increase progressively thereafter. In dogs (Rheingold, 1963) and cats (Schneirla et al., 1963) the period of maternal care has been explicitly divided into three successive intervals characterized by increasing conflict." To cite Schneirla in this context comes close to distortion and clearly points up the predisposition of the sociobiologist to see conflict everywhere. Schneirla described a complex pattern of developing relations including gradual decline of solicitude on the part of the cat and alternating withdrawal and return activity on the part of the kittens leading eventually to mutual independence.

Perhaps the most spectacular instance of the espousal of violence as adaptive behavior concerns infanticide among social animals. The development of this theoretical position is somewhat difficult to trace. In a review of the sociobiology controversy in *Nature*, Robert M. May (1976) describes the behavior of the Serengeti lion referring to studies by Schaller (1972) and Bertram, citing specifically Bertram's article in *The Scientific American* of May 1975. Continuity in a pride of lions resides in the females who may live up to eighteen years. Young males leave the pride at puberty and from time to time the two or three adult males in the pride are driven out and supplanted by two or three nomadic males, likely to be brothers. ". . . soon after taking over a pride," May wrote, "the new males kill the resident cubs . . . since the new males usually share no genes with the existing cubs, the males' inaugural act of infanticide is adaptive from their point of view, although it is certainly not in the females' best interest." This scenario fits perfectly into Hamilton's concept of inclusive fitness and Trivers's notion of parental investment. But this story is not found in Schaller's *The Serengeti Lion* (1972). Schaller describes cub mortality as very high and much of it the result of neglect on the part of the lionesses. Wilson, citing Schaller, says that "cubs are sometimes killed and eaten during territorial disputes" (1975, p. 85). Bertram (1975) says: "On the basis of somewhat scanty data it seems that mortality rate among cubs goes up for about three months after the new males arrive." And, "the available evidence suggests that males on taking over a pride are quite likely to kill the cubs already there."

It is interesting to see how easily this story, which May terms "brutal" when he describes it, has become assimilated as part and parcel of sociobiology. The basic tenets of sociobiology suggest the Hobbesian *bellum omnium contra omnes,* and the way they have been presented has identified the doctrine with a tough-guy *machismo.* When *Time* did a cover story on sociobiology on August 1, 1977, this was the dominant theme.

Behavior similar to that attributed to the Serengeti lion has been claimed for the langurs of India. In the same paragraph in which he mentions the lion, Wilson, citing Sugiyama (1967), says that in a given species of these primates "the young are actually murdered by the usurper." In the January–February 1977 issue of the *American Scientist* there appeared an article by Sarah Blaffer Hrdy entitled "Infanticide as a Primate Reproductive Strategy" with the following summary heading: "Conflict is basic to all creatures that reproduce sexually, because the genotypes, and hence self-interests, of consorts are necessarily non-identical. Infanticide among langurs illustrates an extreme form of this conflict." If this were not enough to place the article in the mainstream of sociobiology, the author acknowledges that valuable comments on the manuscript were made by, among others, R. Trivers.

"Despite the vivid accounts of langur aggression," writes Hrdy, "set down by early naturalists, one of the first steps of modern primatology was to put aside these anecdotes so that the fledgling science of primatology could be laid on a purely factual foundation. By the late 1950s the modern era of primate studies, launched primarily by social scientists, had begun. The early workers were profoundly influenced by current social theory and in particular by the work of Radcliffe-Brown, who believed that any healthy society had to be a 'fundamentally integrated social structure' and that in such a society every class of individuals would have a role to play in the life of the group in order to insure its survival.

"In 1959, Phyllis Jay went out from the University of Chicago to the Indian forest of Orcha and to Kaukori, a village on the heavily cultivated Gangetic plain. Jay found among the North Indian langurs a remarkably peaceful society. . . ."

Hrdy then describes observations by other field workers, including Sugiyama, and by herself consistent with the practice of male infanticide which she insists can be interpreted only on a sociobiological model. In the May–June issue of the same journal there appeared an exchange of letters between Phyllis Jay Dolhinow and Hrdy discussing further the question of male infanticide among langurs. Field studies of animals are difficult and time-consuming. It is impossible to keep animals under continuous observation and critical events have a habit of happening off-stage. Even when the whole of an episode has occurred in full view, there is generally room for varied interpretation, and this will be influenced by one's theoretical model. In this case it is very difficult for one not familiar at first hand with the field work to gain a clear picture of the factual data. The intense emotional involvement of the sociobiologists makes it hard to achieve scientific give and take with them.

The Field of Righteousness

Science consists of hypothetical explanations of natural events deduced from observation and experiment. No scientific theory or principle is immune to competition by another explanation proposed by someone who interprets the observed facts differently. But

ethics is something else. Ethics consists of rules of conduct, imperatives supported by an obligation. Failure to follow the rules constitutes transgression, and the transgressor becomes subject to a
sanction.

Wilson is not content to propose sociobiology as a synthesis of
scientific theory. It is more. It is a means by which, from the observations of animal and human biology, ethics can be found. This
claim is made with different levels of brashness. In an interview
with Nicholas Wade, published in *Science,* Wilson is quoted: "Only
by interpreting the activity of the emotive centers as a biological
adaptation can the meaning of the canons [of morality] be de-
ciphered . . . The time has come for ethics to be removed temporarily from the hands of philosophers and biologized. . . . The
question that science is now in a position to answer is the very
origin and meaning of human values, from which all ethical pronouncements and much of political practice flow."

In a section entitled "The Field of Righteous" in chapter 5 of
Sociobiology, which deals with group selection and altruism, Wilson approaches the subject in a more reverent mood. He invokes an
analogy from Hindu sacred writings, citing an incident in *Bhagvad
Gita* in which Arjuna despairs of attaining peace for the human
spirit. Krishna replies, "For one who is uncontrolled, I agree, the
Rule is hard to attain; but by the obedient spirits who will strive
for it, it may be won by following the proper way." Wilson then
continues: "In the opening chapter of this book, I suggested that
a science of sociobiology, if coupled with neurophysiology, might
transform the insights of ancient religions into a precise account
of the evolutionary origin of ethics and hence explain the reasons
why we make certain moral choices instead of others at particular
times. Whether such understanding will then produce the Rule remains to be seen. For the moment, perhaps it is enough to establish
that a single strong thread does indeed run from the conduct of
termite colonies and turkey brotherhoods to the social behavior
of man."

Wilson gives no precise details on how the human values or the
Rule, the imperatives of righteous behavior, are to be found by
the biologist. These values, since they are adaptive, must have
arisen through the action of natural selection and must, therefore,

be present in the genetic information. They must in some way be self-proclaiming; otherwise they will be subject to variable interpretation and hence unworthy of the capital R.

The search for values is not science. It is theology. It entails very troublesome questions. Who recognizes the Rule? Who judges when it has been transgressed? Who applies the sanction to the transgressor? T. H. Huxley expressed the belief that natural selection itself applied sanctions to those who violated the rules of the game of life. "Yet it is a very plain and elementary truth," he wrote, "that the life, the fortune, and the happiness of every one of us, and, more or less, of those who are connected with us do depend upon our knowing something of the rules of a game more difficult and complicated than chess. It is a game which has been played for untold ages, every man and woman of us being one of the two players in a game of his or her own. The chess-board is the world, the pieces are the phenomena of the universe, the rules of the game are what we call the laws of nature. The player on the other side is hidden from us. We know that his play is always fair, just and patient. But also we know to our own cost that he never overlooks a mistake, or makes the smallest allowance for ignorance. To the man who plays well, the highest stakes are paid with that sort of overflowing generosity with which the strong show delight in strength. And one who plays ill is check-mated—without haste, but without remorse" (1910, p. 58).

Huxley did not purpose to deduce these rules from the hereditary material or to codify them; he was apparently willing to leave the application of sanctions to impersonal forces. Wilson purposes to find and codify the Rule. He identifies it with "ethical pronouncements" and "political practice." It is unlikely that those who see ethics as defined in the genetic material will be willing to leave the punishment for deviant behavior to the impersonal forces of nature. The idea that moral values can be deduced from genetics has within it the terrifying threat of the enforcement of scientifically determined righteousness.

A highly critical review of Dawkins's *The Selfish Gene* by R. C. Lewontin appeared in *Nature,* March 17, 1977. In the issue of May 12, 1977, *Nature* published a letter by W. D. Hamilton denouncing the review. The final paragraph of the letter reads: "Whether the literature of sociobiology reflects an ignorance of the difference between 'properties of sets and properties of their members' or 'confusion between materialism and reductionism' are matters about which I feel little concerned. On the other hand I feel a warmth very far from indifference when I encounter in most of the literature in question signs of a spirit which I share and which I have always assumed is the same as that which motivates scientific enquiry in all its branches. I can most simply express this spirit by calling it a desire to understand and communicate the nature of the world. I find it present in full measure in *The Selfish Gene* and totally absent in Lewontin's review." It gives *me* anxiety to be told that scientific judgment is based on "feeling a warmth" and "encountering signs of a spirit." This sounds to me more like accepting a faith.

And so in perspective, sociobiology is not "The New Synthesis." Rather, it is a shocking attempt to ensnare us all in a set of rules of behavior which are neither to be broken nor questioned because they have been endowed with a pseudo-scientific apotheosis. To provide a source of such rules, a jerry-built doctrine has been compounded of old hat genetics that current research has already rendered obsolete, of sophomorically cynical interpretations of social relations, and of a doctrinaire rejection of the contribution of ontogeny to the behavioral phenotype.

DEFINITIONS

allele   One of two or more alternate forms of a gene occupying a given locus on a particular chromosome. Not all genes containing a given message are coded identically. Two genes coding similar messages which can be shown to be different are alleles. Sometimes each of two alleles of a pair produces a contrasting phenotype— blue or brown eyes, for example. Sometimes the differences are subtle or cryptic as with the blood groups. There may be two, several, or many forms of a given gene. How many there are is related to our techniques for distinguishing them.

heterozygous   Almost all higher organisms possess in every cell two complete sets of genes, one set having been inherited from the mother, the other from the father. Every maternal gene has its paternal homologue and vice versa. If the two genes of any such pair are unlike alleles, with respect to that pair of genes the individual is heterozygous, and may be called a heterozygote.

homozygous An individual who has inherited like alleles of a certain gene from both parents is said to be homozygous for that pair of genes, or, with respect to that pair, to be a homozygote.

locus The position on a given chromosome where a particular gene is found. The DNA of every chromosome exists as one single linear extension. If one measures a certain distance from the end of a chromosome of a given pair, one will find coded there certain functional information—a gene. If one measures an equal distance along the chromosome constituting the other member of that pair, one will find there the same type of functional information. The position of any given gene along the chromosome is its locus (pl. loci). The term is used to identify a relative position with respect to other loci and also to refer to the functional unit found at a given position.

BIBLIOGRAPHY

Bastos, R. N., and H. Aviv 1977 "Globin RNA Precursor Molecules: Biosynthesis and Processing in Erythroid Cells." *Cell, 11*:641–50.

Barlow, H. B. 1975 "Visual Experience and Cortical Development." *Nature, 258*:199–204.

Bertram, B. C. R. 1975 "The Social System of Lions." *Scientific American,* Vol. 232, No. 5 (May 1975), pp. 54–65.

Breathnach, R., J. L. Mandel, and P. Chambon 1977 "Ovalbumin Gene Is Split in Chicken DNA." *Nature, 270*:314–19.

Crick, F. 1979 "Split Genes and RNA Splicing." *Science 204*:264–71.

Crow, J. F., and M. Kimura 1970 *An Introduction to Population Genetics Theory.* New York: Harper and Row.

Darwin, C. *On The Origin of Species* 1859 Facsimile Edition (E. Mayr, ed.). Cambridge: Harvard University Press, 1964.

Dawkins, R. 1976 *The Selfish Gene.* New York: Oxford University Press.

Dobzhansky, T.   1970   *Genetics of the Evolutionary Process.* New York: Columbia University Press.

Flavell, R. A., J. M. Kooter, and E. De Boer   1978   "Analysis of the β-δ-globin Gene Loci in Normal and Hb Lepore DNA: Direct Determination of Gene Linkage and Intergene Distance." *Cell, 15:* 25–41.

Fredrickson, D. S., J. L. Goldstein, and M. S. Brown   1978   "The Familial Hyperlipoproteinemias." *In,* J. B. Stanbury, J. B. Wyngaarden, and D. S. Fredrickson (eds.), *The Metabolic Basis of Inherited Disease.* New York: McGraw-Hill, pp. 604–55.

Haldane, G. B. S.   1955   "Population Genetics." *New Biology* (London), *18:*34.

Hamilton, W. D.   1964   "The Genetical Theory of Social Behavior I & II." *Journal of Theoretical Biology, 7:*1–52.

———   1977   "Letter to *Nature in re* review of *The Selfish Gene.*" *Nature, 267:*102.

Harris, M., and E. O. Wilson   1978   "Encounter: The Envelope and the Twig." *The Sciences, 18:*10–15 & 27–28.

Hrdy, S. B.   1977   "Infanticide as a Primate Reproductive Strategy." *American Scientist, 65:*40–49; and Exchange of Letters with P. J. Dolhinow, same volume, pp. 266–68.

Huxley, T. H.   1910   *Lectures and Lay Sermons.* New York: E. P. Dutton.

Judson, H. F.   1979   *The Eighth Day of Creation.* New York: Simon and Schuster.

Lerner, I. M.   1954   *Genetic Homeostasis.* New York: John Wiley.

Lewontin, R. C.   1974   *The Genetic Basis of Evolutionary Change.* New York: Columbia University Press.

Lorenz, K.   1965   *Evolution and Modification of Behavior.* Chicago: University of Chicago Press.

May, R. M.   1976   "Sociobiology: A New Synthesis and an Old Quarrel." *Nature, 260:*390–92.

Maynard Smith, J.   1978   "The Evolution of Behavior." *Scientific American,* Vol. 239, No. 3 (Sept. 1978), pp. 176–92.

Mayr, E.   1963   *Animal Species and Evolution.* Cambridge: Belknap Press of Harvard University Press.

Parker, G. A.   1978   "Selfish Genes, Evolutionary Games and the Adaptiveness of Behavior." *Nature, 274:*849–55.

Powers, H. W.   1975   "Mountain Bluebirds: Experimental Evidence Against Altruism." *Science, 189:*142–43.

Rheingold, Harriet L.   1963   "Maternal Behaviors in the Dog." *In,* Harriet L. Rheingold (ed.), *Maternal Behaviors in Mammals.* New York: John Wiley, pp. 169–202.

Schaller, G. B. 1972 *The Serengeti Lion: A Study of Preditor-Prey Relations.* Chicago: University of Chicago Press.

Schneirla, T. C., and J. S. Rosenblatt, and Ethel Tobach 1963 "Maternal Behavior in the Cat." *In,* Harriet L. Rheingold (ed.), *Maternal Behavior in Mammals.* New York: John Wiley, pp. 122–68.

Shepher, J. 1971 "Mate-Selection Among Second-Generation Kibbutz Adolescents and Adults: Incest Avoidance and Negative Imprinting." *Archives of Sexual Behavior, 1:*293–307.

Singh, R. S., R. C. Lewontin, and A. A. Felton 1976 "Genetic Heterogeneity within Electrophoretic 'Alleles' of Xanthine Dehydrogenase in *Drosophila pseudoobscura.*" *Genetics, 84:*609–29.

Sociobiology Study Group of Science for the People 1976 "Sociobiology—Another Biological Determinism." *Bioscience, 26:*182–86.

Stent, G. 1978 *Molecular Genetics,* 2nd ed. San Francisco: Freeman.

Stern, C., and E. W. Schaeffer 1943 "On Wild-Type Iso-alleles in *Drosophila melanogaster.*" *Proc. Nat. Acad. Sci. U. S.,* 29:361–67.

Sugiyama, Y. 1967 "Social Organization in Hanuman Langurs." *In,* S. A. Altman (ed.), *Social Communication among Primates.* Chicago: University of Chicago Press.

Trivers, R. L. 1971 "The Evolution of Reciprocal Altruism." *Quarterly Review of Biology, 46:*35–57.

——— 1972 "Parental Investment and Sexual Selection." *In,* B. G. Campbell (ed.), *Sexual Selection and the Descent of Man, 1871–1971.* Chicago: Aldine.

——— 1974 "Parent-Offspring Conflict." *American Zoologist, 14:* 249–64.

Wade, N. 1976 "Sociobiology: Troubled Birth for a New Discipline." *Science, 191:*1151–55.

Wallace, B. 1968 *Topics in Population Genetics.* New York: Norton.

Williams, G. C. 1966 *Adaptation and Natural Selection.* Princeton: Princeton University Press.

Wilson, E. O. 1971 *The Insect Societies.* Cambridge: Belknap Press of Harvard University Press.

——— 1975 *Sociobiology: The New Synthesis.* Cambridge: Belknap Press of Harvard University Press.

——— 1976 "Academic Vigilantism and the Political Significance of Sociobiology." *Bioscience, 26:*183–90.

——— 1978 *On Human Nature.* Cambridge: Harvard University Press.

Wynne-Edwards, V. C. 1962 *Animal Dispersion in Relation to Social Behavior.* Edinburgh: Oliver and Boyd.

GENE-JUGGLING

Genes cannot be selfish or unselfish, any more than atoms can be
jealous, elephants abstract, or biscuits teleological. This should not
need mentioning, but Richard Dawkins's book *The Selfish Gene*
has succeeded in confusing a number of people about it, including
Mr. J. L. Mackie.[1] What Mackie welcomes in Dawkins is a new,
biological-looking kind of support for philosophic egoism. If this
support came from Dawkins's producing important new facts, or
good new interpretations of old facts, about animal life, this could
be very interesting. Dawkins, however, simply has a weakness for
the old game of Brocken-spectre moralizing—the one where the
player strikes attitudes on a peak at sunrise, gazes awe-struck at his
gigantic shadow on the clouds, and reports his observations as cos-
mic truths. He is an uncritical philosophic egoist in the first place,
and merely feeds the egoist assumption into his *a priori* biological
speculations, only rarely glancing at the relevant facts of animal behavior and genetics, and ignoring their failure to support him. There
is nothing empirical about Dawkins. Critics have repeatedly pointed
out that his notions of genetics are unworkable.[2] I shall come to
this point later, but I shall not begin with it, because, damning
though it is, it may seem to some people irrelevant to his main
contention. It is natural for a reader to suppose that his over-

From *Philosophy*, Vol. 54, No. 210 (October 1979). Copyright © Royal Institute of Philosophy. Reprinted by permission of Cambridge University Press.

simplified drama about genes is just a convenient stylistic device, because it seems obvious that the personification of them must be just a metaphor. Indeed he himself sometimes says that it is so. But in fact this personification, in its literal sense, is essential for his whole contention; without it he is bankrupt. His central point is that the emotional nature of man is exclusively self-interested, and he argues this by claiming that all emotional nature is so. Since the emotional nature of animals clearly is not exclusively self-interested, nor based on any long-term calculation at all, he resorts to arguing from speculations about the emotional nature of genes, which he treats as the source and archetype of all emotional nature. This strange convoluted drama must be untwisted before the full force of the objections from genetics can be understood.

Dawkins does toy with egoistic explanations at the more ordinary level as well as with metaphysical "gene" selfishness, although it is not clear why he thinks he needs to. When animals act, as they quite often do, for each other's advantage, Dawkins explains this, where possible, as "reciprocal altruism," that is, not altruism at all but a bargain. It is only when this becomes too obviously unconvincing that he shifts his ground, becoming equally ready to say *either* that the individual is aiming to "increase his own genetic fitness"—i.e. to prosper by having a lot of descendants and relatives —*or* that the real agent is not the individual at all but the personified Gene. This is a mysterious entity riding in the individual and apparently composed of the numerous genes in his cells, which chooses to sacrifice him—and in some sense itself—for the sake of its representatives, with which it somehow identifies, in the descendants who outlive him. I shall discuss the two last alternatives, which are extremely bizarre, later. The first and slightly more respectable idea is the one that seems chiefly to attract Mr. Mackie, because it fits in with traditional egoism. Mackie approvingly cites Dawkins's exposition of it in terms of three imaginary genetic strains in a supposed bird population. They are: Suckers, who help everybody indiscriminately, Cheats, who accept help from everybody and never return it, and Grudgers, who refuse help only to those who have previously refused it to them. These "strategies" are supposed each to be controlled by a single gene, and the help in question is assumed to be essential for survival. In this absurdly abstract and genetically quite impossible situation, Dawkins concludes that Cheats and Grudgers would exterminate Suckers, and Grudgers might well do best of all. Mackie comments with satisfaction that "a grudger is rather like you and me" (p. 410), and reproves Socrates and Christ for supporting Suckers in telling us to return good for evil. "As Dawkins points out," he goes on, "the presence of Suckers endangers the healthy Grudger strategy. . . . This seems to provide fresh support for Nietzsche's view of the deplorable influence of moralities of the Christian type" (p. 464), though he more cheerfully concludes that such moralities are mere words and will have no influence anyway.

Now even if Dawkins's calculations made genetic sense, the only way in which they could provide support for Nietzsche or any other philosophic egoist would be by showing that "reciprocal altruism" or Hobbesian prudential bargaining was the *only* source, or at least far the most persistent and central source, of all animal altruism—in which case we should indeed have good reason to suspect that it was more important than appeared in the human case as well. But the facts of animal life contradict this suggestion entirely. The main source and focus of altruistic behavior in animals is the care of the young, which in most species will certainly never be repaid. Where the young leave home at maturity, parents are lamentably bad Hobbists if they take any notice of their children at all, apart from eating them. Moreover, advanced social species show a great deal of casual and uncalculating friendliness in their lives, and this often proceeds from old to young, from the strong to the weak, even where there is no blood relationship. Calculation about the future is an extreme late-comer in evolution; what forges society are the emotions. Animals are never guided in their lives by any such rigid, simple, games-theory criterion as "did he do it to me last time?" still less "will he be able to do it back?" They can certainly be angry, and to some extent bear grudges. But these events form only one strand among others in the very complex web of social relations which unites them. Within their friendships, mutual help will indeed take place. But the reason will not be the recognition of an insurance premium falling due. It will be liking and affection.

All this is to explain what I mean by calling Dawkins's case absurdly abstract. (It is significant that he could not find a real one). Dawkins supposes the help given to consist in grooming. But then —at least in birds and mammals—the behavior of "sucker" is impossible. Grooming occurs only in social creatures, and occurs there as part of their social bonds. They groom their friends and relations; nobody grooms all comers indiscriminately. This is not a fiddling point. The advantage of being social does not spring from a chance collection of isolated behavior-atoms like hygienic grooming. It is only possible as part of a whole complex way of life in which the outgoing emotions—which egoism denies—constantly work for harmony. (Insects may need to be understood differently, but then neither Mackie nor Dawkins supposes that we are insects.) This disregard of the essential emotional context reappears in Mackie's idea that the undiscriminating "sucker" behavior is one recommended by Socrates and Christ. Neither sage is recorded to have said "be ye equally helpful to everybody." Both, in the passages he means, were talking about behavior to one narrow class of people, with whom we are already linked, namely our enemies, and were talking about it because it really does present appalling problems. The option of jumping on one's enemies' faces whenever possible has always been popular. In spite of its attractions, and in spite of Nietzsche's romantic power-worship, it has proved to have grave drawbacks. Of course charity and forgiveness have their drawbacks too, especially if they are unintelligently practiced. As Mackie rightly says, there are problems about reconciling them with justice, and justice too has its roots in our emotional nature. There are real conflicts here as both Socrates and Christ realized. But since they are real they cannot be much helped by a dashing gesture towards Nietzsche.

In dealing with these problems Dawkins's grossly simplified and distorted scheme is no use at all. Suckers do not exist. A blank, automatic, undiscriminating disposition to help everyone in sight would be pathological in any animal. In a human being, it would certainly not pass as charity or forgiveness but simply as loopiness. No doubt this, along with the equal dottiness of "cheat," is what gives Mackie the impression that, by comparison "a Grudger is rather like you and me." Being a shade less simple he certainly is *more* so, but the difference is trifling. We find him slightly less hard to believe in because he seems to show signs of being able to distinguish between friends, enemies, and strangers. And we, like any other social animal, regard this as a paramount condition of normal life. But the signs are deceptive, because the Grudger is supposed to view as enemies all those who have ever failed to return his help in the past, and as friends all those who have returned it. This principle, on which a man's employer would usually be his best friend and his children always his enemies, is unknown in the animal world. Altruism is transitive long before it is reciprocal. No one who has heard of evolution has any business to suppose, as Hobbes excusably did, that calculating prudence is the root of all social behavior. Now that we know how complex the social life of other species can become when their intelligence does not make calculation possible, we know that there is no such single root. Ethological comparison strongly confirms what an unprejudiced view of the human scene has always suggested, that motivation is complex. There is no short cut to understanding it. In each case we have to look at the detailed evidence.

The particular case which Mackie raises of the way in which the injured treat their injurers is a good instance of the surprising complexity which we find when we do this. In the species most like our own, lasting resentment after injuries is by no means a prominent or important motive. In some cases, of course, immediate fighting is possible, but prolonged grudge-bearing is rare and trivial. Jane Goodall notes with interest how in her chimps the usual effect of an injury is something very different—a distressed approach to the aggressor with a demand for reconciliation. What seems to be most noticed is not the injury itself, but the failure of the social bond:

> A chimpanzee, after being threatened or attacked by a superior, may follow the aggressor, screaming and crouching to the ground or holding out his hand. He is, in fact, begging a reassuring touch from the other. Sometimes he will not relax until he has been touched or patted, kissed or embraced (*In the Shadow of Man*, p. 221).
>
> While a male chimpanzee is quick to threaten or attack a subordinate, he is usually equally quick to calm his victim with a touch, a pat on the back, an embrace of reassurance. And Flo,

after Mike's vicious attack, and even while her hand dripped blood
where she had scraped it against a rock, had hurried after Mike,
screaming in her hoarse voice, until he turned. Then as she approached him, crouched low in apprehension, he had patted her
again and again on her head, and as she quietened, had given her a
final reassurance by leaning forward to press his lips on her brow
(p. 114).

As she points out, this reaction makes it possible to resume the relationship as though the injury had never taken place. (A community of retentive "grudgers" would by contrast be a terribly in-
secure one; no lapses would be tolerated.) She rightly remarks,
too, that small human children do the same thing. It is only for
adult human beings, with their much stronger powers of memory,
imagination, and foresight, that this simple reaction becomes impossible. The whole problem then takes on another dimension of
complexity.

Altogether, I know of no evidence from the behavior of other
species to suggest that prolonged grudge-bearing is anywhere a
powerful motive. It can hardly, then, be an important root of justice. By contrast, readiness to fight back *immediately* in case of injury certainly is such a root. Actual animal-watching shows that this
tendency is nothing like as strong or as common as has often been
imagined. Still, there are plenty of situations where it does occur,
usually either between individuals of roughly equal status, or between strangers, or on occasions of exceptional outrage. But to
grow into the emotional raw material of justice, this capacity for
instant retribution needs another element. It has to become vi-
carious; that is, altruistic. And it does so. Dominant animals often
do attack middle-ranking ones who are bullying their inferiors, and
may take the inferiors under their permanent protection. But this
too is an outgrowth of parental protectiveness; again it presents
the problem of altruism. In fact, the account that Mill gave of the
matter is a fair one, provided that it is understood, not as an analysis of the notion of justice, but as an account of its psychological
origins:

> The sentiment of justice appears to me to be the animal desire
> to repel or retaliate a hurt or damage to oneself, or to those with

whom one sympathizes, widened so as to include all persons, by
the human capacity of enlarged sympathy, and the human conception of intelligent self-interest (*Utilitarianism,* Ch. 5).

But the fact that there are "those with whom one sympathizes" at
all is ruinous to simple-minded egoism. "The human capacity of enlarged sympathy" certainly makes the point still more pressing, but
the simplest case of parental care in animals already presents it in
a damning form.

This persistent difficulty in reducing parents to the egoist pattern
is just the kind of thing which makes Dawkins's typical readers—
people with vaguely egoist leanings about individual human psychology—willing to follow him in losing touch with the observed
facts of motivation altogether and taking off for the empyrean with
the Gene. Dawkins, however, does not even start from those facts.
He draws all his material from "sociobiological" evolutionists such
as W. D. Hamilton, Edward O. Wilson, and John Maynard Smith
who are not directly interested in individual psychology at all. (Incidentally, his pages are virgin of originality except for a single
suggestion which I shall discuss in my last section.) These evolutionists' main business has been to show how conduct which does
not benefit the agent can survive in evolution by benefiting his
kin; they have worked out the arithmetic of "kin-selection." This
way of thinking actually makes any dependence on individual
selfishness as a motive unnecessary, and the term "selfish" should
not appear in their writings. For some reason, however, they are
still devoted to it. Even the least romantic of them, W. D. Hamilton, has a paper called "Geometry for the Selfish Herd," and Wilson takes enormous pains to show that a great range of obviously
uncalculated altruistic human behavior, such as impulsive rescuing,
is *really* bargaining, and therefore concealed selfishness.[3] They
show a strong and unexamined tendency to assume both that individual motivation must actually, despite appearances, be selfish,
and that it makes sense to talk of entities other than individuals as
being selfish. R. L. Trivers, closely followed by Dawkins, has in-
flated this bad habit into a mythology. Before examining it, however, it is worth while asking why dogmatic egoism exerts this
powerful pull. At the quite unthinking level, of course, it has two

great attractions, both of which it shares with Hedonism—its great
apparent simplifying power, and its swashbuckling style. But anyone who is so far intrigued by these as to begin applying it in detail
quickly finds that the facts are too complicated for it. The first advantage is illusory. The second, though very influential in accounting for Dawkins's success, cannot be the only factor determining
his mentors; two other more serious reasons come in.

The first is an error that has always dogged this controversy,
namely, an unrealistic notion of altruism. People define altruistic
behavior negatively, as activity which while helping others does
nothing for the agent, which he himself does not at all want, or
which is necessarily to his disadvantage. This negative conception
seems to destroy the possibility of motivation towards it. The word
however means something positive. The act is done *for* the benefit
of another. Helping him is the *aim,* one's own feelings are the in-
ducement; one's own disadvantage forms no part of the idea. It is
mere confusion to suppose that satisfaction taken in it, or its happening to turn out useful to one, make it a selfish act. Bishop Butler long ago nailed this error:

> If, because every particular affection is a man's own, and the
> pleasure arising from its gratification his own pleasure, such
> particular affection must be called self-love, according to this way
> of speaking, no creature whatever can possibly act but merely from
> self-love. But then, this is not the language of mankind; or if it
> were, we should want words to express the difference between the
> principles of an action, proceeding from cool consideration that
> it will be to my advantage; and an action, suppose of revenge, or
> of friendship, by which a man runs upon certain ruin, to do evil
> or good to another (Sermon XI, Sec. 7).

Altruism, in fact, is not a fantastic concept, but a descriptive one
with a use to distinguish some existing motives from others. Besides this familiar difficulty, however, the evolutionary context
adds another, newer and more confusing factor. In natural selection, many are born but few survive for long. We call this "competition," and the metaphor at once suggests the specific *motive* of
contentiousness. As we begin to grasp the scale of the phenomenon,

the strength of the motive involved seems to grow. Before Darwin drew attention to it, nobody, probably, realized how many must die early as the necessary condition of the life and development of a few. My present business is not with the problems of theology but with the confused way in which people have persistently attributed to individual creatures the motives which seem needed in an imaginary being who might actually understand, and will, this whole process in which he is involved. Darwin, just because he was an exceptionally humane man, was shaken by what he found, and often used terms like "war" and "remorseless struggle." Being a realistic naturalist, however, he would never have made the mistake of supposing that mice and mushrooms, pigs and pampas-grass were actually busy on unscrupulous plots to destroy each other, still less that minute scraps of their cell-tissues were so occupied. Only quite advanced creatures are sufficiently conscious of each other's existence to "compete" in the full sense of the word—to know what they are about and have the appropriate motives. (Even human beings do not usually do so.) Predators, as their expressive movements show, do not regard their prey with anger or cruelty or as a fellow-creature at all, but just as meat. Remorse could not enter into the matter, so "remorselessness" in the true sense of determined callousness cannot either. For the same reason, the milder notion of "selfishness" is equally out of place. Among social birds and mammals we might use it, though hesitantly, to describe an individual who constantly grabbed more than his share. But for non-social creatures we could not use it so, since no question of shares arises among them. Similarly, a robin driving intruders off his territory cannot be supposed to weigh up their claims, predict their subsequent starvation, and decide in his own favour. He is not selfish; he just wants the place clear. One cannot speak even of "unthinking selfishness" in beings incapable of the thought in question. Most selective competition does not require competitive motives, nor any sort of motive involving calculation of consequences, and much of it requires altruistic ones. Absolutely none of it below the human level can proceed from dynastic ambition. Moreover, dynastic ambition is not selfishness, but a particular complex human motive which may well conflict with self-interest. The further down the scale of creatures we go, the more obvious all this becomes. Nobody attributes selfish planning to a paramecium. What, then, can Dawkins mean by attributing it to a gene?

Doing his best for Dawkins, Mackie ignores this point, but it cannot be ignored; as its title implies, the book depends on it. Dawkins brings in gene motivation because his account of individual motivation is a total failure; in fact, he switches from one to the other with bewildering speed every time he gets into a difficulty. About individual motivation he would like to be an egoist, but the facts of ethology prevent it. He wants to relate the workings of natural selection in a simple and satisfying way to those of motivation by finding a single universal motive, and there is no such motive. Having picked on selfishness for this role, he personifies genes in order to find an owner for it. It may indeed seem that he must just be speaking metaphorically, as he sometimes claims. But the trouble about these admissions is that Dawkins seems to have studied under B. F. Skinner the useful art of open, manly self-contradiction, of freely admitting a point that destroys one's whole position and then going on exactly as before. When ruin stares him in the face, he withdraws into talk of metaphors, but he goes on afterwards as if the literal interpretation still stood. For instance, on p. 95 of *The Selfish Gene:*

> If we allow ourselves the license of talking about genes as if they had conscious aims, always reassuring ourselves that we could translate our sloppy language back into respectable terms if we wanted to, we can ask the question, what is a single selfish gene trying to do? . . . "It" is a distributed agency. . . . A gene might be able to assist *replicas* of itself which are sitting in other bodies. . . .

In short, because a gene cannot perpetuate *itself* but only likenesses of itself, the language of selfishness is so crashingly wrong that even Dawkins sees he will have to hide it under the table for a bit, even from people who were willing to make a pet of his bogus entity. But this by no means makes him go back and alter the flat, unfigurative assertions which are everywhere essential to the book's argument or modify its opening manifesto:

> [This book] is not science fiction; it is science. Cliché or not, "stranger than fiction" expresses exactly how I feel about the truth.

> We are survival machines—robot vehicles blindly programmed to preserve the selfish molecules known as genes. This is a truth which still fills me with astonishment (p. x).

Not a word of caution about metaphors follows. On p. 210, Dawkins has the gall to write, "Throughout this book, I have emphasized that we must not think of genes as conscious, purposeful agents." These disavowals do occur now and then, but, like the paternosters of Mafia agents, they have no force against his practice of habitually relying on the literal sense. On p. 48, too, he takes a very different line. Resisting people who might say that he has "an excessively gene-centered view of evolution," he makes the quite proper and moderate reply that study of genetic causes is useful. Then, evidently concluding that genes have been shown to be the only reality, he suddenly adds:

> At times, gene language gets a bit tedious, and for brevity and vividness we shall lapse into metaphor. But we shall always keep a sceptical eye on our metaphors, to make sure they can be translated back into gene language if necessary.

This seems to mean that not only the talk of conscious motives, but also all talk of whole organisms and their behavior, is only a metaphorical way of describing the behavior of genes. Anyone who can talk like this has a deeply confused view of metaphor, and a few words on this topic seem called for.

To understand how metaphors can properly be used in scientific writing, we must get straight a fundamental point about the relation between metaphors and models. Every metaphor suggests a model; indeed, a model is itself a metaphor, *but one which has been carefully pruned.* Certain branches of it are safe; others are not, and it is the first business of somebody who proposes a new model to make this distinction clear. Once this is done, the unusable parts of the original metaphor must be sharply avoided; it is no longer legitimate to use them simply as stylistic devices. For instance, the familiar model of *mechanisms* in biology has long ago been pruned of its original implication that a mechanism needs an inventor or maker. Anyone writing about a "biological mechanism" knows that he must keep such inventors out of his explanation. He must somehow manage to use the language of purpose and adaptation without this reference; figurative speculations about the inventor's character and history will damage and confuse his reasoning. He may want to do theology, but if so, he must do it explicitly, not by loosely extending the language of "mechanisms."

Just so Dawkins, in officially discussing the merely physical action of genes, constantly uses the language of conscious motive and depends entirely on it to create the impression that he is in a position to say anything about human psychology. Calling genes selfish is indeed a metaphor. Whatever may be deemed to be the usable part of this metaphor, which might fit it to become a model, everyone will agree that the attribution of conscious motive belongs to the unusable part. Yet that attribution is the only thing which makes it possible for him to move from saying "genes are selfish" to saying "people are selfish."

If anyone has any doubt about this, it may be best dealt with by moving on to examine the supposedly safer branches, to ask "what then, ignoring the figurative flourishes, is the literal sense which the metaphor is there to convey?" Shorn of its beams, it turns out to be a point about the ultimate "unit of selection":

> The fundamental unit of selection, and therefore of self-interest, is not the species, nor the group, nor even, strictly, the individual. It is the gene, the unit of heredity (p. 12, cf. p. 42).

> Genetically speaking, individuals and groups are like clouds in the sky or dust-storms in the desert. They are temporary aggregations or federations. They are not stable through evolutionary time [whereas the gene] does not grow senile. . . . It leaps from body to body in its own way and for its own ends. . . . The genes are the immortals (p. 36).

The suggestion seems to be that, in order to understand the behavior of larger units or "temporary aggregations," all that we need is to understand the behavior of genes. This looks like a simple recommendation to go and do some genetics. Dawkins, however, is no geneticist, and when we ask for further information on how genes do behave, he invariably returns to what was supposed to be merely a metaphor:

> Can we think of any *universal* qualities which we would expect
> to find in all good (i.e. long-lived) genes? . . . There might be
> several such universal properties, but there is one which is particu-
> larly relevant to this book; at the gene level, altruism must be
> bad and selfishness good. . . . Genes are competing directly with
> their alleles for survival. . . . The gene is the basic unit of selfish-
> ness (pp. 38–39).

The reason why he cannot get off this subject is not that he knows
no genetics, but that all the genetics which he or anyone else
knows is solidly opposed to his notion of genes as independent
units, only contingently connected, and locked in constant inter-
necine competition, a war of all against all. (In spite of some
words in the last quotation, he cannot really mean that it is just
war between each gene and its own alleles; this would allow cooperation over the rest of the field and destroy his case entirely.)
What he needs is a "prisoners' dilemma" situation, in which each
unit operates alone, and does it in the same way whatever the others
may do. What he has got is a situation of the utmost causal complexity, in which genes probably always vary their workings according to context, always depend on each other, and in many cases may
produce a totally different effect when different "modifier" genes
accompany them.

It is time to turn to the genetic realities. As I have suggested,
Dawkins's crude, cheap, blurred genetics is not just an expository
device. It is the kingpin of his crude, cheap, blurred psychology.
For selection to work as he suggests by direct competition between
individual genes, the whole of behavior would have to be divisible
into units of action inherited separately and each governed by a
single gene. Something like his simple sucker/cheat model would
have to be adequate right across the board. One gene must govern
each "strategy" if their "interests" are supposed to be always in
competition. To convince us that this is so, Dawkins brings up
once more the case of Rothenbuhler's Hygienic Bees, creatures
which have been appearing in suspicious isolation as a stage army
in all such arguments for some time, and, as if it were both well
proven and typical, he airily adds, "If I speak, for example, of a
hypothetical gene 'for saving companions from drowning' and you

find such a concept incredible, remember the story of the hygienic bees" (p. 66). Actually, not only does the bees' case stand alone, but it is certainly not proven. To show that even the simple behavior it involves is really governed by only two genes would take something like seventy generations of outbreeding experiments to ensure that the effects described are not due to the close linkage of genes at a whole series of adjacent loci, and even this would not show that these genes affected nothing else.[4] (By Dawkins's account, Rothenbuhler has studied two generations.) Those are the standards to which geneticists work. Genetics is that complicated. It is so because—as is well known—genes are essentially co-operative; they are linked together in the most complex and hierarchical ways and affect each other's working to an incalculable extent. The idea of a one-one correlation is not genetics at all. As Dobzhansky put it, tracing the history of his subject in 1962:

> The original conception of simple unit-characters had to be given up when it was discovered that the visible traits of organisms are mostly conditioned by the interaction of many genes and most genes have pleiotropic, or manifold, effects on many traits. . . . Although geneticists no longer speak of unit characters, others continue to do so. . . . The academic lag goes far to explain why so many social scientists are repelled by the idea that intelligence, abilities or aptitudes may be conditioned by heredity (*Mankind Evolving*, p. 33).

This refers to work done before 1920. Since that time, the emphasis on interdepedence among genes has steadily grown. In his offhand way, Dawkins acknowledges some of this in Chapters 3 and 4. But this in no way embarrasses him when he writes of "the grudger gene" (p. 199) nor when he repeatedly assumes in those same chapters that each gene is a quite independent force wielding enormous individual influence. Thus, in considering how sexual reproduction arose, he writes that this would indeed be hard to understand in terms of advantage to the individual or even the increase of his posterity:

> But the paradox seems less paradoxical if we follow the argument of this book, and treat the individual as a survival machine built

by a short-lived confederation of long-lived genes. "Efficiency" from the whole individual's point of view is then seen to be irrelevant. *Sexuality versus non-sexuality will be regarded as an attribute under single-gene control, just like blue eyes versus brown eyes.* A gene "for" sexuality manipulates all the other genes for its own selfish ends (p. 47, my italics).[5]

Occurring in a student's genetics essay, the italicized sentence would just be a bad mistake. It cannot be turned into something else here by the metaphorical context, because this point is not part of the metaphor; it is what the metaphor is meant to convey as literal fact. The context does, of course, make a difference, because what in a student would be simple ignorance is here being used to bail out an unworkable thesis. The same open disregard for consistency surrounds the questions of the gene's credentials as a unit. Its unity and permanence are, as the quotations just made show, supposed to be its great merits. Dawkins however cheerfully acknowledges what is well known; that the word "gene" is used in various senses by geneticists for varying sections along the DNA, and that none of them is immortal. In fact the word may be used to indicate different lengths of DNA within the chromosome depending whether a unit of mutation, function, or recombination is being referred to. These are so far different that Dawkins's clanger is like that of someone analysing language, who insists that we must find its fundamental elements, but talks as if it did not matter whether we take those elements to be letters, words, or sentences. Aware of trouble here, he hastily adopts a general definition for "gene" which he attributes (rather surprisingly and without reference) to George Williams. A gene is now defined as "any portion of chromosomal material which potentially lasts for enough generations to serve as a unit of natural selection" (p. 30). This, he claims with relief, is the end of his search for "the fundamental unit of natural selection, and therefore the fundamental unit of self-interest. What I have now done is to *define* the gene in such a way that I cannot help being right." That is: in physical terms, what he says is tautological and meaningless; he might be talking about any section of the DNA, though obscurely. In psychological terms, it is both meaningless and absurd, since he has linked the notion of self-interest quite gratuitously to a kind of subject for which it can make no sense at all. The only possible unit of self-interest is a self, and there are no selves in the DNA.

When the mountains of metaphor are removed, in fact, what we find is not so much a mouse as a mare's nest, namely the project of finding a unit which will serve for every kind of calculation involved in understanding evolution; a "fundamental unit" at a deep level which will displace, and not just supplement, all serious reference to individuals, groups, kin, and species, and which (for some unexplained reason) will be also the unit of selfishness or self-interest. Dawkins is not the only person to be impressed by the idea of a universal unit, but it is vacuous. To see how vacuous, we might ask the parallel question, "what is the fundamental unit of economics?" A coin? If so, how large and of what country? A single worker? A factory? A complete market exchange? A minimal investor? For various purposes and from different angles, we might need to count any of these things. The decision which to count, and how finely to divide them, would depend entirely on the particular problem which we wanted to solve, and for most purposes we would refer to all of them, and would rightly not expect to have to reduce one to another. The reason that Dawkins gives for electing genes to this strange position in evolution is that they are less changeable than the entities of which form part. But as far as this goes, physical particles are in a stronger position still. Dawkins sometimes does toy with this thought, calling them too "selfish replicators"; why stop at genes? The reason can only be that our understanding of genes does a special job in explaining evolution. This is true, but, since genes are not on view, it is a limited job, entirely dependent on a direct understanding of the more obvious entities in their own terms. Moreover, physical particles can exist without organisms; genes cannot. They survive only if their owner belongs to a species, and one which has not fallen below the critical frequency for further breeding. Members of a population within a species probably have as many as 70–80 per cent of their genes in common (ignoring "neutral alleles" whose results (allozymes) make no difference and are therefore "invisible" to selection).[6] And these genes are hierarchically linked in such a way that any serious disturbance of the group will not give rise to a viable organism at all. (This is why hybrids are usually sterile.) Genes are units indeed for some purposes of calculation, but they are not independent, privateering units. If a gene *were* a conscious planner, it would have to reckon its interests as including those of a mass of other genes on which it is dependent, as well as all such genes in all possible mates for its owner's descendants, and all necessary ancestors for those mates —in short, everything needed for the gene pool—in short, since any gene pool can fall into trouble, everything needed for the whole species, and indeed for the eco-system. No biological unit can be both "fundamental" in the sense of lasting, and also independent. But this is no tragedy, since there is no sort of need for such a unit. Physics itself no longer looks, as it used to, for "atoms" in the strict sense of unsplittable units, permanent and unchangeable billiard-balls, at the end of its analysis. There is no point at all in other sciences dressing up in its old clothes and inventing such units.

There is however a perfectly good controversy carried on among evolutionists about the "unit of selection," one dealing with a real but much more limited issue. We ask: *what* is it that natural selection selects? Now there is an obvious and perhaps conclusive sense in which we must answer "individuals." Organisms are born and die as wholes; each does not directly involve another, but it does involve all its parts. The notion of "group selection," however, was invented to account for the fact that some ways of behaving seem adapted rather to preserve the group than the individual. (This thought arose not so much about altruistic behavior as about population mechanisms which look like devices to stabilize the size of a group.) But the phrase "group-*selection*" is confused, because what is selected ought to be items out of a set. And it does not normally happen that many distinct groups compete without mixing. Instead there is usually gene-flow between them, and group-stabilizing characters spread throughout the species. "Group-selection" is a bad term if it is taken to mean something parallel and *alternative* to individual selection. All the same, the point raised is a real one, and draws attention to a confusion in the notion of "selection" itself. Organisms are selected *as* individuals, but what are they selected *for?* The term "select" leads people to hope for a simple, positive answer to this question, a single, isolable purpose. We would like to say, "just as an employer choosing workers selects simply the ones who will maximize profits, so evolutionary pressures select simply those who will maximize something specific like their own life-span." But neither employers nor pressures can really act so simple-mindedly. The idea of an "economic man" whose *sole* aim is to maximize profits cannot be made coherent. This is not only because, if he is a man as well as being economic, he will be moved by non-economic considerations like not wanting to go to jail or work himself to death. It is because we do not know, and economics cannot tell us, *at what time* the profits are to be counted. Security for next year, or for some such slice of the future, normally counts as a condition to be satisfied before profits start to be reckoned; indeed, the notion of "profits" is normally understood against a background of this condition and many others, such as not murdering all possible rivals. But in principle one could decide to aim at absolute maximization in six months followed by suicide, or alternatively, as misers do, to live in penury with a view to maximizing at the end of the longest possible life-span. Quite different policies would follow these decisions. Puzzles remarkably like this infest the attempt to find a single aim for natural selection. Sociobiological thinkers are inclined to hope they can solve them by substituting "maximum genetic fitness" for "maximum life-span" as the aim of selection. But this is mere word-spinning. "Maximum genetic fitness" means having as many surviving relatives as possible, and this simply *is* "being selected"— it is not the aim or condition of it. Just as with economics, the degree of "success" achieved will seem to vary with the time when one decides to do the audit. Changes long after an individual's death can bring his hitherto unwelcome genes into sudden demand; webbed feet or a silent habit become necessary in new circumstances. But they might not have done, and it is idle to say "then he was fitter than we supposed"; after all, we might have to reverse the judgment again later on.

It is probably necessary, for evolution as for economics, to think not of one single aim, but of a number which converge, and particularly to notice a number of negative conditions which must be met. No sensible economist supposes that his subject lays bare the ultimate structure of human life and reveals its deep determining purpose. Evolution, however, is a much larger and more complex thing than human life, less likely still to yield to formulation in such simple terms. Even if we confine ourselves to asking what is needed for an individual to be "selected"—to survive and leave descendants—we shall not find one goal which he has to reach, but rather a great many disasters which he has to avoid. His own qualities can only account for some of them. Some are outside anyone's control, some—a great many in social species—lie in the control of con-specifics. Mutual aid and protection can be quite essential to him, and they are more often transitive than reciprocal. Because they largely occur among kin, this point has been expressed by talking of "kin-selection," which means the development of kin-profiting behavior by the selective advantage which it gives to those kin-groups which practice it. This is a reasonable idea, though again it is not actually an *alternative* to individual selection. As with larger groups, the picture is not one of isolated kin-groups competing, but of protective behavior spreading through the advantage it confers. Since kin-groups are normally not exclusive, this spread will eventually go beyond them. "Kin" in fact is not the name of a super-entity which *replaces* individuals in the selection process, but a pointer to the necessarily social character of some behavior. This social character can have various ranges. Parental care helps chiefly one's kin. The mobbing of predators helps chiefly one's group. Migration and colonization may help chiefly one's species. In all these kinds of cases, the reason why the behavior can develop is that it helps to build up the supportive background needed by all individuals rather than directly helping the agent.

Thus the notions of kin- and group-selection each have a point, but it is one which can be expressed compatibly with the obvious truth expressed by the notion of individual selection. Real empirical issues remain, about just how the mechanisms involved work, both socially and physiologically, but a blank clash of polarized views is unnecessary. Gene-selection, however, which Dawkins puts forward as the winning candidate for this somewhat unreal race, is a much more obscure idea. Because of the genetic complications I have mentioned, it is hard to give it any meaning at all. As Stephen Jay Gould sensibly puts it:

No matter how much power Dawkins wishes to assign to genes, there is one thing that he cannot give them—direct visibility to natural selection. Selection simply cannot see genes and pick among them directly. It must use bodies as an intermediary. . . . Bodies cannot be atomized into parts, each constructed by a single gene. . . . Parts are not translated genes, and selection doesn't even work directly on parts. It accepts or rejects entire organisms. . . . The image of individual genes, plotting the course of their own survival, bears little relation to developmental genetics as we understand it "Caring Groups and Selfish Genes" (*Natural History,* Vol. 86, Dec. 1977).

Why, finally, does this matter? There are many aspects of it which I cannot go into now, and I concentrate on the moral consequences which Dawkins and Mackie draw. Egoism, when it is not just vacuous, is a moral doctrine. It has, as Mackie sees, always a practical point to urge. Aristotle used it to tell us to attend to our own personal and intellectual development. Hobbes used it to urge citizens to treat their government as accountable to them generally, and particularly to make them resist religious wars. Nietzsche, non-political and often surprisingly close to Aristotle, did on his egoist days preach self-sufficiency and self-fulfillment as a counterblast to the self-forgetful and self-despising elements in Christianity. But he is only a part-time egoist. Any attempts to use him as a signpost here would, as usual, be frustrated by his equal readiness to denounce bourgeois caution and exalt suicidal courage, or "love of the remotest." He hated prudent bargaining. His egoism is confused, too, by contributions from his personal terror of love and human contact. Still, against the wilder excesses of Christianity he certainly had a point, and he was able to make it without any reference to genes. Is there any way in which reference to genes could become relevant to disputes about it? Dawkins makes the connection as follows:

The argument of this book is that we, and all other animals, are machines created by our genes. Like successful Chicago gangsters, our genes have survived, in some cases for millions of years, in a highly competitive world. This entitles us to expect certain qualities in our genes. I shall argue that a predominant quality to be expected in our genes is ruthless selfishness. . . . Let us try to teach generosity and altruism, because *we are born selfish* (pp. 2–3, my italics).

He contends, that is, that the appearance of "a limited form of altruism at the level of individual animals" including ourselves, is only a deceptive phantom. The underlying reality, as he often says, is not any other individual motivation either, but the selfishness of the genes. Yet he just as often talks as if this established that the individual motivation *were* different from what it appears to be—as here, *"we* are born selfish." His thought seems to be that individual motivation is only an expression of some profounder, metaphysical motivation, which he attributes to genes, and is bound therefore to represent it. And he has arrived at his notion of gene-motivation by dramatizing the notion of competition. Even as drama, this fancy is gratuitous. All that can be known about our genes from the fact that they have survived is that they are strong. If people insist on personification, the right parallel would no doubt be with a situation in which a number of travellers had, independently, crossed a terrible desert. It might happen that in doing so they had unknowingly often removed resources which would have saved the lives of others—but this could tell us nothing about their characters unless they had known that they were doing so, and scraps of nuclear tissue are incapable of knowledge. We could be sure only that such travellers were strong, and to make a parallel here we must examine the concept of gene "strength." This strength is not an abstract quality, but is relative to the strains imposed at the time. The fact that people have survived so far shows only that they have had the genetic equipment to meet the challenges they have so far encountered. Human pugnacity had its place in this equipment. But since people are now moving into a phase of existence when pugnacity itself becomes one of the main dangers to be faced, new selective pressures are beginning to operate. In this situation telling people that they are *essentially* Chicago gangsters is not just false and confused, but monstrously irresponsible. It can only mean that their feeble efforts to behave more decently are futile, that their conduct will amount to the same whatever they do, that their own and other people's apparently more decent feelings are false and hypocritical. On the other hand, to tell them (what is quite different) that they have actually no motives at all and no control over their actions, that they live in a permanent state of post-hypnotic suggestion, helpless pawns in the hands of powers over whom they have no influence, is melodramatic and incoherent fatalism. The unlucky thing is that people enjoy fatalism, partly because it promotes bad faith and excuse-making, partly because the melodrama has a sado-masochistic appeal—an appeal which gets stronger the nastier the powers in question are supposed to be.

Dawkins, however, claims innocence of all this. He says he is merely issuing a warning that we had better *resist* our genes and "upset their designs."

> Be warned that if you wish, as I do, to build a society in which individuals co-operate generously and unselfishly towards a common good, you can expect little help from biological nature. . . . Let us understand what our own selfish genes are up to, because we may then at least have the chance to upset their designs . . . p. 3).

He does not explain who the "we" are that have somehow so far escaped being pre-formed by these all-powerful forces as to be able to turn against them. He does not even raise the question how we are supposed to conceive the idea of "building a society in which individuals co-operate generously and unselfishly towards a common good," if there were no kindly and generous feelings in our emotional make-up. He does however see some difficulty in accounting for the diversities of human conduct. This so far disturbs him that he produces for once an idea of his own, not derived from Trivers, Hamilton, Wilson, or anybody else—the idea that cultural evolution is a process on its own, taking place in units called *memes* (short for mimemes):

> Examples of memes are tunes, ideas, catch-phrases, clothes fashions, ways of making pots or of building arches. Just as genes propagate themselves in the gene pool by leaping from body to body via sperms or eggs, so memes propagate themselves in the meme pool by leaping from brain to brain via a process which, in the broad sense, can be called imitation (p. 206).

These memes, equally with genes, are selfish and ruthless:

> When we look at the evolution of cultural traits, and at their sur-
> vival value, we must be clear whose survival we are talking about.
> . . . A cultural trait may have evolved in the way that it has,
> simply because it is *advantageous to itself*. . . . Once the genes
> have provided their survival machines with brains that are cap-
> able of rapid imitation, the memes will automatically take over.
> We do not even have to posit a genetic advantage in imitation
> (pp. 214–15).

So apparently, if we want to study (say) dances, we should stop
asking what dances do for people and should ask only what they
do for themselves. We shall no longer ask to what particular human tastes and needs they appeal, how people use them, how they
are related to the other satisfactions of life, what feelings they express or what needs cause people to change them. Instead, presumably, we shall ask why dances, if they wanted a host, decided
to parasitize people rather then elephants or octopuses. This is
not an easy question to handle for dances, but it will still be harder
for scientific theories. Dawkins explicitly includes them as memes,
so that the proper way to enquire about them seems to be, not to
investigate their truth or any other advantage which they might
have for the people using them, but to study the use they make of
people. Here, to be frank, Dawkins blathers, and no wonder. The
idea of memes is meant to save human uniqueness, to avoid producing the sense of insult which readers often feel on being told
that their traits are inherited, and which they have a right to feel
ten times more strongly after the account which Dawkins has
given of inherited traits. But it is still an explanation of the only
kind which (apparently) Dawkins can conceive, namely a metaphysical one in terms of autonomous, parasitical, non-human entities. Again it is unrelated to the facts, and on top of that this
time it fails still more obviously and resoundingly in the job of
providing "units." A meme is a meant to be "a unit of cultural
transmission, or a unit of imitation." In the case of genes, Dawkins has insisted very firmly on the permanence, distinctness, and
separability needed for such units, and because the general public

does not realize that genes do not have it, he has more or less got by. In the case of "memes" the simplest observer can see that no such standards can be met. Consequently, even if—absurdly—imitation were the essence of culture, it could not have units and the whole conception falls to the ground. Besides this, of course, the theory not only fails to give a proper, workable account of human freedom but sets up another, apparently impenetrable, barrier in the way of supposing that we are free at all. No wonder, then, that Dawkins hurries past his half-finished meme-construction to advise us, in peroration, to save ourselves from "the worst excesses of the blind replicators" including memes. We are to do this partly by improved calculations of self-interest, but also, he says, partly by "deliberately cultivating and nurturing pure, disinterested altruism—something that has no place in nature, something that has never existed before in the whole history of the world. We are built as gene machines and cultured as meme machines, but we have the power to turn against our creators." Why it should be imagined that Dawkins and his disciples, beginning this enterprise now, could succeed when everyone else in recorded and unrecorded history who has tried it has managed only to become infested by memes (including scientific theories), does not emerge. Nor is it clear whether Mr. Mackie is going to welcome this new enterprise.

Over memes there is, of course, a nightmare possibility of developing Dawkin's case. In a sufficiently depressed mood, a psychologist might really feel moved to describe the history of human thought in terms of its progressive infestation by conscious, self-interested, parasitical *bad ideas*. For the time, that might seem to him the only way of explaining the confusion he sees, the chronic waste of human speculative intelligence, the contentiousness, the showing-off, the neglect of obvious facts. In this project, he might well find his most convincing examples in theories of motivation, and specially in those (like Dawkins's) which simplify it by reduction and trade on fatalism. This topic is, of all important human enquiries, perhaps the hardest to approach impartially, the most prone to distortion both by oversimplification and bad faith. Modern specialization, too, has made it even more vulnerable to bad theories by dividing the critics who should provide immunity

against them. There is now no safer occupation than talking bad
science to philosophers, except talking bad philosophy to scientists.
Should we then (he might wonder) resign ourselves to enduring
all such manifestations, including *The Selfish Gene,* as impreg-
nable alien life-forms, a kind of mental bacillus against which no
antigen can ever be developed? Emerging finally from his bad
mood, however, he would find strength to resist this idea. Entities
(he would remind himself) ought after all not to be multiplied
beyond necessity. Spooks should not be encouraged; less super-
stitious explanations are not hard to find. Slapdash egoism is not
really a very puzzling phenomenon. It is a natural expression of
people's lazy-minded vanity, an armchair game of cops-and-rob-
bers which saves them the trouble of real enquiry and flatters their
self-esteem. No non-human intervention is needed to account for
it; it is a commonplace, understandable disorder of human development, like obesity or fallen arches. It is no subject for science fiction; ordinary care and attention are enough to remedy it.[7]

Note In the context of this book I had better make plain what
does not emerge in this article: that I am not by any means opposed to every aspect of sociobiology, but only to some of its excesses. In my book *Beast and Man* (Cornell University Press,
1978) I discussed the larger issues involved and drew conclusions
which I still hold, namely that
1. Innate tendencies are extremely important in the formation of
human behavior.
2. Recognition of these innate tendencies is not in the least inimi-
cal to human freedom. On the contrary, if there were no such
tendencies, the concept of freedom would be unintelligible.
3. Careful and informed comparisons with the behavior of other
species can illuminate human behavior. They enable it to be
placed in its evolutionary context, and are an entirely proper and
useful contribution to our understanding of it.
   I therefore find much of the investigation contemplated by sociobiologists perfectly inoffensive. However, that makes it all the
more necessary to dissociate these legitimate, familiar, general
ways of thinking from certain quite unnecessary excrescences
which belong peculiarly to sociobiology: the mindless fatalism;

the bad genetics; the casual, distorted, slapdash treatment of the psychology of motive. In a way, Dawkins performs a service by specializing in these faults and inflating them to the point of caricature. It seems only necessary to read him to grasp that *this* can't be the way to understand and investigate the innate element in human life. A reader who sees this and loses interest in sociobiology seems to me to have the right reactions. His next step should be to turn to the work of Eibl-Eibesfeldt, where he will find the subject properly handled, and also—if he hasn't already done so—read Konrad Lorenz's book *Behind the Mirror: A Search for a Natural History of Human Knowledge* (Harcourt Brace Jovanovich 1977). This is a far more interesting and seminal work, which entirely avoids the crude reductionism that repeatedly erupts to spoil Wilson's work, let alone Dawkins's.

My own further thoughts on this controversy may be found in an article called "Rival Fatalisms: The Hollowness of the Sociobiology Debate," which also appears in this volume.

A plague, to some extent, on both your houses. Both badly need to be set in order.

NOTES

1. J. L. Mackie, "The Law of the Jungle," *Philosophy* 53 (October 1978).
2. The attempt which he has eventually made to answer some of these criticisms may be read in *Zeitschrift für Tierpsychologie* 47 (1978), 61–76. Apart from some minor disputes, it simply intensifies the conceptual blunders which I discuss here. Dawkins always answers opponents who point out that "genes" as scientists normally conceive them cannot possibly play the role which he assigns to them by retreating still further from the facts to a more general metaphysical position where "genes" are classed as "replicators." Unless he either learns to do metaphysics or retreats out of sight entirely, this is not going to do him any good.
3. See *Sociobiology* (Harvard University Press, 1975), 120. He adds, however, "Human behavior abounds with reciprocal altruism consistent with genetic theory, *but animal behavior seems to be almost devoid of it.*" He accounts for this (as I do) by the lack of calcula-

tion in animals, but seems not to see that, since these "animals" are the subjects we are dealing with for almost the whole of evolution, any "genetic theory" inconsistent with their capacities will have to be revised. Dawkins, in his "Grudger" story, ignores Wilson's reasoning here, as he does most other things that do not suit him.

4. For an example of such work fully carried through, see Kyriakou, Burnet, and Connolly on heterozygote advantage in the mating behavior of *Drosophila* ("The behavioral basis of over-dominance in competitive mating success at the *ebony* locus in *Drosophila melanogaster*"). *Animal Behavior* 27 (1979).

5. Contrast with this confident and startling pronouncement a typical passage from the Preface to John Maynard Smith's thoughtful book *The Evolution of Sex* (Cambridge University Press, 1976): "I am under no illusion that I have solved all the problems that I raise. Indeed, on the most fundamental questions—the nature of the forces responsible for the maintenance of sexual reproduction and genetic recombination—my mind is not made up. On sex, the relative importance of group and individual selection is not easy to decide. . . . It has struck me while writing that the crucial evidence is often missing, simply because the theoretical issues have not been clearly stated."

6. See R. S. Singh, R. C. Lewontin, and A. A. Felton on "Genetic Heterogeneity within Electrophoretic 'Alleles' of Xanthine Dehydrogenase in *Drosophila pseudoobscura*," *Genetics* 84 (1976), 609–29.

7. I would like to acknowledge invaluable help over the scientific side of this paper, given by my colleague, Dr. A. L. Panchen of the Zoology Department of the University of Newcastle.

S. A. BARNETT

BIOLOGICAL DETERMINISM AND THE TASMANIAN NATIVE HEN

For some people, man is made in God's image; for others, *Homo sapiens* is an aberrant primate and, like all other species, a product of evolution by natural selection. There are those who accept both notions. If we take the biological view, we may try to explain human behavior as an evolutionary "strategy" that enabled our ancestors to survive and multiply in competition with others: hypotheses on how animal social behavior evolved are now often proposed, and similar hypotheses are being applied to ourselves. For convenience, I refer to those who put forward these notions as sociobiologists.

Such neodarwinian arguments are often based on unproved and unacknowledged presumptions. When they are applied to man they have social implications which are likely to be disregarded or glossed over. The disregard sometimes seems to reflect the writers' lack of acquaintance with relevant aspects of knowledge outside their own fields. The massive accumulation of detail in the natural sciences, especially biology, has forced us to be narrow specialists. Nonetheless, every specialist has a "philosophy" or set of attitudes about the world he lives in. And, since the human condition presents many baffling questions, solutions of intractable problems are attempted by applying simple concepts based on special study (*cf.* Barnett, 1979).

Based in part on passages from *Modern Ethology: The Science of Animal Behavior* (to be published by Oxford University Press).

The first part of this essay is therefore on a special branch of biology: it examines current speculations about the evolution of animal social behavior. In the second part, I turn to some consequences of speculating in the same way about man.

A. *Animals*

NEODARWINISM

In evolutionary accounts of social interactions a central concept is Darwinian fitness. In biology, fitness is measured by the contribution an individual makes to the next generation. This point of view is represented by Samuel Butler's epigram, that a chicken is an egg's way of ensuring production of more eggs. The theory of natural selection can be stated with more rigor than Butler attempted, but only with difficulty. It was indeed never made precise by Darwin (1873, 1901) or by his immediate successors. The best-known logical difficulty is presented by the phrase, used but not originated by Darwin, "the survival of the fittest": if the fitness of a genotype is defined in terms of its chances of survival, the expression becomes empty—the survival of those that survive.

Darwin and his successors evaded this issue by using literary imagery (Young, 1971). The very word "selection" implies a selector; it corresponds to Darwin's frequent references to the changes undergone by animals and plants under domestication—that is, selection by men (on which he wrote a lengthy treatise). But it is misleading to speak of natural selection as if it were something like a stockbreeder. Nonetheless Darwin, in many passages, refers to nature (a feminine deity) as the agent of selection; sometimes she is called "a power." The notion of a *struggle* for survival is also prominent in his writings.

The theory of natural selection requires that within a species there is genetical variation. We now know that such variation is generally found in natural populations. Second, the genetical variation must result in differences of fitness among the different types. But these notions have been current for millennia (Haldane, 1959): by themselves, they point only to the tendency for some types to be eliminated; and, if the atypical forms are the ones to be weeded out, the result is stability. Darwin's originality rested in the notion that natural selection could also produce change.

We now have many examples of such change on a small time-scale. Indeed, some are common knowledge. Widespread use of insecticides has resulted in the appearance of populations immune to their effects. Antibiotics work miraculously at first, but resistant bacterial strains soon appear. These and other well-observed micro-evolutionary changes are consequences of environmental disturbance due to human action. The human species is not exempt from such influences. The eugenic movement has been justly criticized as founded on ignorance and class prejudice (Allen, 1975; Kamin, 1974); but, in so far as it has a scientific basis, it is an attempt to apply the principles of genetics to the improvement of human populations.

By an irony, one of the best-documented examples of eugenic improvement due to human action has resulted from a purely environmental measure. Sickle-cell anemia is an extremely unwelcome recessive condition which has for long been prevalent in West African and other populations with a high incidence of subtertian malaria (Allison, 1969). The way to reduce the incidence of the unwanted sickle-cell gene is to get rid of the malaria. Where this has been done, the gene is weeded out (another metaphor) by natural selection. Hence, not only are human populations subject to neodarwinian effects, but an understanding of such processes also has important practical implications. The extent to which other environmental improvements have had eugenic effects is, however, not known.

THE TASMANIAN NATIVE HEN AND OTHER STORIES
It is one thing to have examples of change due to natural selection, another to account for the diversity we see around us. Much effort has gone into explaining, in neodarwinian terms, the variety of animal mating patterns (Clutton-Brock & Harvey, 1978; Krebs & Davies, 1978). As an example, Maynard Smith and Ridpath (1972) have discussed wife-sharing by male Tasmanian native "hens" (*Tribonyx mortierii*). This species has more males than females; and the permanent breeding groups commonly consist of one female and two males. A neodarwinian therefore asks why one

male does not drive away the other. Conflict does occur among these birds: they are not natural pacifists. And in other quite closely related species, not only is there conflict among males but there is monogamy: indeed, most birds are monogamous (Lack, 1968). Moreover, there are many species in which only some individuals of each population succeed in mating.

On the face of it, a male's tolerance of another, in a *ménage à trois,* puts him—in a neodarwinian sense—at a disadvantage: it tends to reduce the proportion of his genes that will be represented in the next generation. The explanation offered is that the two males in each group are brothers. Hence each will share many genes in common. The net effect of polyandry in this species is therefore held to fit the neodarwinian framework: although wife-sharing halves (on average) the male's chance of passing on genes to his own offspring, the net genetical effect (given an excess of males) is favorable. (There is no cogent explanation of the unusual sex ratio.)

On such an hypothesis, we have here an example of kin selection. This notion, which we owe especially to Maynard Smith (1964) has been reviewed by Clutton-Brock and Harvey (1978). Nature is not necessarily "red in tooth and claw," for cooperative behavior is widespread in the animal kingdom: members of the same species are not in continuous conflict, but often combine in such a way that the survival of two or more individuals, each potential rivals (or of their offspring) is made more likely. The problem presented by such conduct arises from our presumption that all characteristics are subject to natural selection. If this is correct, conduct that helps conspecifics should be eliminated whenever it tends to reduce the fitness of the helper: that is, when it diminishes the chances of survival of the helper's genes. The obvious exception is when the beneficiaries are the helper's offspring; but, as implied in the example above, the notion of kin selection can be applied to relationships between any individuals that have a substantial number of genes in common. Hence it is supposed that cooperative behavior becomes increasingly likely with increasingly close genetical relationship.

A notable set of examples comes from species, including some bees and birds, of which only some individuals breed, but these are

assisted by others (Brown, 1978). Among the birds, the green
woodhoopoe, *Phoeniculus purpureus,* is an example (Ligon &
Ligon, 1978). Flocks may include up to sixteen adults, but each
has only one breeding pair. The others help in the care of the young
by feeding them. Such relationships have been described for quite
a variety of species, including the Florida scrub jay, *Aphelocoma
coerulescens* (Woolfenden, 1975), and the superb blue wren of
Australia, *Malurus cyaneus* (Rowley, 1965). Helpers may not
only feed young, but also contribute to nest-building and to defense
against predators. To an ordinary human observer, such conduct
seems most agreeable and attractive; but from a neodarwinian
standpoint it may be thought anomalous: Are the helpers close relatives of those helped?

Before examining this argument further, I take two more examples of reproductive patterns. (There are many others.) Each presents a neodarwinian enigma. There is first the work of Robertson
(1972) on *Labroides dimidiatus.* This is a cleaner fish, that is, a
species which feeds on the ectoparasites of other fish. On the Great
Barrier Reef these fish form groups, of which each consists of a
male with up to six mature and some immature females. If a male
dies the oldest female of the group rapidly changes sex: conduct
toward other females alters in a few hours, and the whole change
takes only two to four days.

There are other such protogynous hermaphrodites with more
females than males, but only a few. What is the function, that is,
survival value, of such a system? We can only speculate. Robertson
points out that each male fertilizes the eggs of a number of females; hence, of all the members of a group, his genotype contributes the most to later generations. But, it may be supposed, the
longest-lived female, that ends up as a male, is likely to be the best
adapted to the immediate conditions; if not, she would have failed
to achieve seniority. If she differs genetically from the others, her
genotype is the best one for the environment. Such a system is then
economical in its production of males, but also perhaps allows
rapid genetical adaptation to environmental demands.

Yet, if these suppositions are correct, it may reasonably be asked
why such systems are so rare. To answer, we should have to speculate further on the advantages of alternatives. But if we do this, we

are gravelled for lack of matter: there is no information on which to base a comparison between the system of *Labroides,* on the one hand, and that of more conventional fish, on the other. All we can do is suggest how this unusual life history *might* have aided survival.

The last case is that of the red deer, *Cervus elaphus,* studied by Lincoln (1972) and by Lincoln and Guinness (1973). In the breeding season some stags assemble a harem of as many as twenty hinds, and drive away other males. The successful males have magnificent antlers. If these structures are lost, by accident or experiment, the effect is a decline in status: stags deprived of antlers have no access to hinds during the rut. Probably, antlers are important as visible signals, and contribute to success in courtship. But there is an anomaly: a few stags (hummels) fail to grow antlers, yet succeed in breeding. Somehow, it seems, they adapt their conduct to their condition, and use displays in which the antlers are not involved.

There are other species of which the males are dimorphic. The question for us now is how such a system could have evolved. Gadgil (1972) suggests that in every dimorphic species the difference between the two forms is genetically determined, and that the forms represent alternative strategies. One strategy is to develop powerful or conspicuous structures; these may be supposed to enhance reproductive success, but at a cost in the energy needed to develop and to carry them. The alternative strategy is to economize on energy, at the cost of inferiority in signaling powers.

Gadgil, like some other model-builders, states that more information is needed on natural populations. But such findings can rarely, if ever, answer the question: did the features observed evolve in the manner proposed by the model? If evidence appears that seems not to fit the model, subsidiary hypotheses can, of course, always be proposed. For example, members of the same species seem often to cooperate regardless of the closeness of their relationship. Now suppose that an animal slightly endangers itself, and in doing so greatly improves the chances of survival of another. Such conduct could have survival value for the actor, if there were a high probability of the beneficiary returning the "favor" on some later occasion (Trivers, 1971). If, however, one wishes to find out whether such reciprocal relationships exist, there are considerable difficulties. One is that an apparent case of general reciprocity is difficult to distinguish from cooperation among kin.

Many social and reproductive systems have now been analysed in sociobiological terms (see, for example, West Eberhard, 1975). On all such analyses, an obvious comment may be made: they show that what actually happens is also theoretically possible (*cf.* Lewontin, 1979). Perhaps this is reassuring. Maynard Smith (1978) has put the point less harshly: "I believe that the main role of models in evolutionary biology is to help us to see whether, in particular cases, the proposed causes (. . . selection pressures) are sufficient to account for the observed results."

ECOLOGY HERE AND NOW

In fact, however, models or hypotheses about the workings of natural selection can do more than that. In ethology the principal use of the evolutionary framework today is to help us to describe and to classify the vast variety of animal behavior patterns, and (with more difficulty) to relate these patterns to mode of life or ecological niche. In addition, evolutionary models have implications for the population genetics of existing species: they can lead to hypotheses testable by study of extant forms. Such studies do not, in themselves, logically require any evolutionary presumptions. We can investigate the ecology and reproductive strategies of animal species which have different habitats, social systems, and feeding habits; and we can try to establish how these are related. Such study requires detailed information, on many aspects of the lives of many species, which is only now beginning to be acquired; but at least the information, unlike the evolutionary processes that have acted in the past, is accessible.

As an example I take a review by Orians (1969). The title is *On the Evolution of Mating Systems in Birds and Mammals,* and there is much discussion of how the conduct of extant forms could have evolved; but the empirical content of the argument concerns what animals do today, and how this is related to the conditions in which they live. The birds have among their mating systems not only polyandry, such as that of the Tasmanian hen, but also polygyny. Orians asks what neodarwinian advantage or survival value is conferred by the possession of several wives. (It is, of course, equally appropriate to ask about the value of monogamy.)

Orians mentions first the hypothesis that polygyny is a means of regulating population growth. Such a proposal seems to require that the males can respond, not merely to the density of the population, but to the level of mating activity: their libido should disappear when their group has achieved a certain amount of breeding (an analogue of the zero population growth urged on ourselves). Orians writes: "This theory would be best tested by direct demonstration of the processes that are postulated." (The same comment could justly be made of any theory proposed about events accessible to inquiry.) Empirical questions include the following. (i) Do females of polygynous species fail to mate because, after the males have achieved a certain level of sexual activity, the males themselves refuse further coitus or the females reject their advances? (ii) Do some (peripheral) males become unresponsive to females once the population reaches a certain stage?

In fact, what findings there are do not fit these proposals. In particular, when polygynous species have been closely studied, no unmated females have been found, nor any evidence of refusal by males to mate with receptive females.

An alternative hypothesis accepts that polygyny must always tend to be advantageous to the successful males, and asks: in what conditions is it also advantageous to females? To answer, it must first be presumed that males vary in reproductive capacity. A female's chances of rearing young are then aided if she has the ability and opportunity to choose one among a number of males. She should then (one might suppose) prefer a male who would give her undivided attention; and indeed, most birds are monogamous. But suppose an environment has a varied and patchy distribution of the resources (food, nesting places) needed for breeding. Males establish their territories in spring, and advertise for females by singing or other displays. It might then be advantageous for a female to choose a male with an especially well-furnished territory, even if he has already attracted other females.

The argument leads to a "prediction"—that polygyny should occur where resources are patchy, as in marshlands. Other such relationships between mating patterns and ecological niche may be inferred, and sometimes the inferred relationships are actually found; but, even among the birds, which are the most studied group, there are many exceptions (Selander, 1972).

Outside the birds, one of the problems is the existence of monogamy and paternal care among some species of mammals (Kleiman, 1977). We "expect" monogamy among birds, because a male can do everything for the young except produce and lay the eggs; but only a female mammal can lactate: a male mammal is in general essential only for insemination, and so—according to one argument—should be promiscuous and spread his semen as widely as possible. As Kleiman shows, there is no convincing explanation of the way in which the different mating systems are distributed among species. Why, for example, should gibbons (Hylobatidae) be monogamous, but the closely related chimpanzee, *Pan,* and gorilla, *Gorilla,* be promiscuous?

The reviews cited bring out both the empirical and the inferential content of sociobiological models. (i) Each species has an ecological niche or mode of life: it lives in a particular habitat, eats only certain food, and so on. Devising a useful classification of niches is an important task. (ii) Social (and other) behavior is closely related to niche. Hence (iii) there should be correlations between type of niche and type of social system (including mating pattern). (iv) "Predictions" are then made in the form: if you look in environment E_1 you will find social system of type S_1; but in E_2 you will find S_2. (v) When such correlations are found, it is presumed that they tell us something of how natural selection produced existing forms. (vi) When correlations are not found, or are weak, either new models are needed, or subsidiary hypotheses must be devised to "save the phenomena." The method of (v) and (vi) is *post hoc,* not predictive. Despite much work, both in the computing room and in the field, the scope of this method is still a matter of debate.

B. *Man*

BIOALTRUISM AND HUMAN EVOLUTION

The preceding account of "sociobiology" does not do full justice to the ingenuity of writers such as Hamilton (1964), Trivers (1972), and others (see Clutton-Brock & Harvey, 1978). It is also in terms which may seem strange to some readers, for I have said nothing of selfish genes, nepotism, bullying, spite, or even parental invest-ment. Neodarwinist writings now conventionally present an orgy of ethical and economic anthropomorphisms. The most conspicuous example is the misuse of altruism. This term, in its customary meaning, refers to a person's intentions: a lexical definition is: *regard for others as a principle of action.* In contrast, we have this: "In biological terms, an altruistic act . . . decreases the fitness, that is, the chances of survival and reproduction of the actor, while increasing the fitness of a conspecific" (Krebs & May, 1976). Here there is no reference to intention, but only to the consequences of what is done. The moral concept disappears. To avoid confusion, for the biological concept I use the term *bioaltruism* (Barnett, 1979).

The importance of this distinction arises from current specula-tions—or, sometimes, dogmatic assertions—about human evolu-tion, in which the argument on natural selection, outlined above, is applied to our own conduct. Hence "ethics and ethical philosophers" are to be explained "at all depths" by the fact that "they evolved by natural selection" (Wilson, 1975). Evidently, being an "ethical philosopher" can contribute to the chances of one's genotype being represented in later generations. And so we have a picture, that of the selfish family man, which is superficially not unlike that of Thomas Hobbes, the seventeenth-century philosopher, for whom the central feature of human conduct was egoism, not cooperation. (But for Hobbes the selfishness was rational, not an automatism.) This is one of several current biological models of man, each based on a simple concept derived from the researches of specialists. The other evolutionary models now prominent present mankind as a product and an exponent of ruthless violence. In its most extreme form our species is represented first as a hunter, that is, a predator; hence we kill individuals of other species to feed ourselves. Second, we are tool-makers. Third, we are ineradicably violent. Fourth, our tools are weapons directed not only against other species but also against our own kind. And, finally, we are cannibals (Dart, 1954).

The lurid fantasies of Dart and his successors have been fully exposed by Montagu (1976): there is no good evidence that our ancestors were persistently violent toward each other, let alone cannibals; and infanticide and cannibalism, when they do occur, are ritual rather than nutritional practices (Dickeman, 1975). Even the notion of Paleolithic man as primarily a hunter is questionable. Most modern pre-agricultural groups are predominantly plant-eating, and are appropriately called gatherer-hunters (see, for example, Lee & DeVore, 1968). The handbag or shoulder sack was perhaps a more significant invention than the spear or the bow (*cf.* Clark, 1976).

There are therefore alternative "new models" of human evolution. The possibilities are exemplified by Jolly (1970). Some of the hominid forms that preceded man lived on savannah. Today baboons, *Papio,* live on grassy plains, and occupy an ecological niche in which there is little competition with other species; like ourselves and our predecessors, they are ground-living. Jolly suggests that, during the Tertiary, our ancestors began to depend largely on the grass seeds available in open savannah, and so too adopted a niche in which there was little competition. Unfortunately, seed-eating lacks the drama of the chase. Moreover, Jolly ruins his chances of becoming a public figure by refusing to be dogmatic: he accepts that his proposal is only an hypothesis that can hardly be tested.

LESS THAN KIN, BUT MORE THAN KIND

One of the reasons why man as a species is difficult to pin down, or to force into a single mold, is that he has no one ecological niche. Our Pleistocene ancestors may have lived in open plains, but they contrived to have descendants capable of extraordinary diversity. Even before agriculture was invented, man had spread over most of the earth's surface: gatherer-hunters can cope with the Arctic, tropical forest, hot deserts, and oceanic islands, as well as regions where now populations are densest.

It is a truism that we find, similarly, immense cultural diversity. Sexual relationships are of central importance in sociobiological theory. Sexual practices that are or have been held to be natural and normal in human communities include monogamy, polyandry, polygyny, and promiscuity; and in some societies homosexual conduct is or has been taken for granted. Hence we cannot apply to our whole species the type of sociobiological analysis given to other species, most of which are confined to a single mode of life.

How should one try to explain the variety of human marital, extramarital, and nonmarital practices in neodarwinian terms? The obvious, common-sense reply is that it cannot be done. The extreme cases are exclusive homosexuality or total celibacy. To deal with such phenomena, the method used to explain, or to explain away, awkward species-typical patterns of animals has nonetheless been applied also to the human species. "The homosexual members of primitive societies may have functioned as helpers . . . with a special efficiency in assisting close relatives" (Wilson, 1975). They may; or they may not. Such statements have no explanatory value.

Wilson also writes: "Spite is commonplace in human societies, undoubtedly because human beings are keenly aware of their own bloodlines and have the intelligence to plot intrigue." The following sentence might equally be held to be "undoubtedly" true: "Spite is universally condemned in human societies, because human beings are morally and intellectually equipped to reject it." We are free to choose which of these two attitudes we prefer. The proponents of the selfish model of man are, however, unlikely to be moved by references to the actual conduct of human beings. Their arguments are based on deductions from neodarwinian premises, not on the findings of empirical study. But in science and in practical affairs, observations of the real world are all-important.

For systematic observations of human conduct we have to turn to anthropology. Sahlins (1977) reviews this field: he finds "not a single system of marriage, postmarital residence, family organisation, interpersonal kinship, or common descent" that conforms with sociobiological "principles of kin selection." Some of the diverse phenomena covered by his brief survey include infanticide as an accepted and regular practice, the adoption of the children of dead enemies, and an explicit rejection of the importance of "blood" (genetical) ties.

A similar disregard of human diversity is found in the treatment of what zoologists sometimes, rather loosely, call the pair bond. To quote Wilson (1975) again, "The traits of physical attraction are . . . fixed in nature. They include the pubic hair of both sexes and the protuberant breasts and buttocks of women." Such features are held to determine "the close marriage bonds that are basic to human social life." But standards of sexual attractiveness vary greatly between societies and even between individuals. Moreover, marital relationships are much influenced by other aspects of social custom, including property relationships and other economic factors (Sahlins, 1977).

ALTERNATIVE DETERMINISMS

Sociobiological theory, then, interprets man as a creature of fixed impulsions determined by the genes and by evolution. It is instructive to compare it with another popular biological determinism—one that goes a little toward recognizing man as *Homo sapiens,* not merely *H. egoisticus* (selfish) or *H. pugnax* (aggressive). Our species is above all capable of learning. In this century an energetically asked question has been: What determines what we learn or, more precisely, what habits we form?

The application to our own species of simple ideas on habit formation is associated with the work of two men, I. P. Pavlov (1928) and B. F. Skinner. The concept of conditional reflexes had an impact far outside psychology, partly as a result of Pavlov's own writings. For example, he referred to religion as "socially conditioned reflexes." Perhaps this expression was intended more as a provocation than a serious contribution to social psychology. But an indication of the impact of Pavlov's work during much of this century is found in the fact that the expression "conditioned reflex" is generally familiar. Moreover, in conventional textbooks of human or mammalian physiology, usually addressed to medical students, we find, in volume after volume, after the chapters on the nervous system a simplified account of conditional reflexes. No other information is given on brain function or on the physiology of behavior; and there is no acknowledgment of the inapplicability of most current neurophysiological knowledge to normal or abnormal human behavior.

The concept of the conditional reflex, despite its historical importance, has severe limitations, even when applied to animal behavior. Yet Pavlov's work also influenced the writings of at least one celebrated philosopher. Bertrand Russell (1958) regarded CRs as "characteristic of human intelligence"; though in an earlier work (1927) he points to some of the limitations of CRs as explanations of how habits develop. These limitations appear also in Shaw's

The Adventures of the Black Girl in Her Search for God. Shaw describes how his heroine meets Pavlov.

> ". . . If I give you a clip on the knee you will wag your ankle."
> "I will also give you a clip with my knobkerry; so dont do it" said the black girl.
> "For scientific purposes it is necessary to inhibit such secondary and apparently irrelevant reflexes by tying the subject down" said the professor.

As Shaw implies, an animal strapped to a table and secreting saliva (or changing its heart rate) is not adequate as a model of man, or even of how a human being learns things.

The other prominent attempt to apply "learning theory" to man is that of Skinner (1953, 1957, 1973). His experimental technique consists of putting an animal in a box, and recording how often it operates a key or a lever. The animal can do little else except move about restlessly. The experimenter varies the consequences of key-pecking or bar-pressing, and so influences the rate at which these acts are performed. The act may (or may not) reward the animal (with food and so forth), or punish it (usually with an electric shock). Such an experimental procedure disregards all the complex phenomena of exploratory behavior (curiosity); and Skinner has nothing to say about the ability of animals (let alone man) to produce *novel* adaptive acts in response to need. Similarly, he ignores imitation. For criticisms see Chomsky (1959), Peters (1974), and Flew (1978).

A fundamental deficiency is in Skinner's concept of reinforcement. In effect, positive reinforcement (or reward) is defined in terms of the influence of external events on the subject's habits: whatever increases habit strength is called positively reinforcing. Hence we have the statement that a pianist learns to play scales smoothly because "smoothly played scales are reinforcing." This is Skinner's answer to the question, why does a pianist (try to) play scales smoothly? Accordingly, in this system, "what is reinforcing" is equivalent to whatever is actually done, and Skinner's statement about reinforcement contains no objective information beyond the statement that the behavior described actually occurred: the pianist plays scales smoothly because he plays scales smoothly. Assertions on punishment, or aversive stimuli, are similarly vacuous: "Punishable behavior can be minimized by creating circumstances in which it is not likely to occur."

Skinner himself refers to a critical comment: "The traditional view [of man] supports Hamlet's exclamation, 'how like a god!', [while] Pavlov, the behavioral scientist, emphasized 'how like a dog!'" For Skinner, the "behaviorist" position is "a step forward"; Skinner does not, however, discuss the possibility that human beings do not like being treated as dogs. Perhaps they find it aversive.

Skinner's theories have led to no fruitful program of research on the complexities of human behavior. Indeed, their inadequacies, even when applied to other species, are well known. The attempt to apply them to human societies, like those of Pavlov and his followers and of the sociobiologists, seems to reflect a desire for simple explanations of phenomena which cannot in fact be simply explained.

LANGUAGE, TRADITION AND TEACHING

Of these phenomena, the human ability to communicate by a complex language represents particularly clearly the gulf between ourselves and other species. The achievements of chimpanzees, *Pan troglodytes,* in learning a gestural language, though interesting (Gardner & Gardner, 1975; Premack, 1976), do not bridge this gulf. Young chimpanzees do not spontaneously learn speech of any kind, as human children do. More important, they do not, however highly trained, learn to debate either empirical matters or questions of logic. At the risk of hubris, I state that no chimpanzee can do what I am doing as I write these words—not even by gestures.

I can do this because I inherit a tradition of such debate, traceable for millennia. ("Inherit" is here used in its social sense.) Tradition depends on *teaching*—a distinctive feature of man (Barnett, 1977). By "teaching" I mean behavior designed to alter the conduct of another, *and persisted in until the other reaches a certain standard of performance.* Although many mammalian species teach their young in some sense, no others teach skills in this way. Moreover, teaching (so defined) seems to be universal in human communities: in particular, children teach younger children.

Such behavior, though often described, has been strangely neglected as a subject in its own right. Perhaps this is partly because of its extreme complexity. It certainly fits no simple sociobiological framework. It may be regarded as faintly analogous to the performance of "helpers" such as those we find among bees and birds. But, unlike "helping" by animals, it entails the exercise of the distinctively human traits of self-consciousness, empathy, altruism (in its primary sense), and the use of language.

Among these items Slobodkin (1978) has emphasized the importance of the idea of the self, or self-consciousness, as a feature of human life. Although chimpanzees, too, are perhaps to some extent self-conscious (Gallup, 1970), they lack a language in which to articulate self-criticism. I am now referring to the ability to observe the self as something comparable with objects or with other persons. Such an ability allows us to use language to talk about ourselves in elaborate ways. We can debate what we observe, imagine future possibilities, and decide on which features of human conduct to promote, which to reject.

These aspects of humanity are species-typical. But unlike the stereotyped, species-typical conduct of animals, of which I give examples above, they allow both cultural and individual diversity, on a scale not represented elsewhere. Once again, we see that, while other animals are each adapted to a particular ecological niche, we are instead adaptable to a great range of environments. Hence human society has, as a general feature, the capacity to change rapidly, in a way unmatched by even our nearest zoological relatives. A major peculiarity of our species is in our changeable cultures, centered on language.

Other species change only slowly, as a result of the action of natural selection on genetically determined variation, provided, in the long run, by mutation. The analogue of mutation in our social evolution (if there is one) is the development of a new idea or an invention. Medawar (1959) uses the term "exosomatic heredity" for the social process by which novel (and other) notions are transmitted. Transmission by tradition is, however, quite different from the biological process by which genes are passed on. There is no parallel to the fundamental distinction between genotype and phenotype. Exosomatic units are "inherited" (if at all), intact—like an item of property: no dilution of their effect is enforced by sexual reproduction.

An idea or invention can, of course, influence biological fitness. No doubt the discovery of fire, around a million years ago, favored the survival of the discoverer and his or her immediate kin. The same applies to the development of agriculture about ten thousand years ago. It was certainly followed by a vast increase in the world population. With time, the number of examples has increased exponentially: in this century alone we have, among many others, the discovery of vitamins, antibiotics, and the oral contraceptives. Even the last can, in some circumstances, increase fitness.

INDETERMINISM

Determinist arguments seem to imply that all such developments are an inevitable consequence of natural selection, or of other processes which are independent of human choice or intention: those who believe that they have some freedom to choose must then be deluded. But nearly all adults, in practice, reject this position (Lucas, 1970). We say that a person *ought* to take some action, and so imply that the person has a choice. We may also say, "serve him right" and imply that a person has a moral responsibility for his actions. These presumptions about our conduct are among the axioms on which human society is based. The work of scientists, including sociobiologists, like that of other human beings, is founded on the prior belief that we have some freedom of action. According to Arieti (1972), the only people who consistently reject the possibility that they are free to make choices are certain schizophrenics. Such people are held to be insane, and their views are generally rejected. This is not, in itself, evidence against determinism; but it illustrates the axiomatic character of our presumptions about the will.

Sociobiologists and other modern determinists do not often deal explicitly with the general questions raised by their views. On the face of it, a scientist may seem *obliged* to reject freedom and responsibility. His objective, whenever he goes beyond the mere description of particular items, is usually to predict what will happen: in the attempt to make correct predictions he uses causal relationships that he or others have discovered or hypothesized. That powerful method is taken for granted as valid in the everyday work of science. What we are now faced with is the possibility that it has limitations. The apparent dilemma can be resolved if we treat the scientists' method, not as all-embracing, but as a procedure employed wherever it is effective: we stop short when we come to a self-contradiction or absurdity.

To try to force all phenomena into the framework of the natural sciences is to disregard other kinds of knowledge, including the presumptions on which the sciences rest. If we are automata, then our belief that we (or some of us) can create something new is false: the moral principles of a Confucius, the theories of a Newton, the works of a Shakespeare or a Mozart, are comparable to the printout of an automatic typewriter programmed to respond in predetermined ways to certain inputs. But we reject that position as we go about our ordinary affairs. Indeed, there are formal arguments which demonstrate that complete determinism is logically untenable (reviewed by Lucas, 1970; MacKay, 1960). Hence, to avoid inconsistency, we are obliged to recognize that the powerful methods of science are not all-powerful: there is much that they can predict, or hope to predict; but they should not be expected to cope with everything.

THE SOCIAL DIMENSION

Sociobiology and other determinisms (applied to man) may be looked at in at least three ways. First, in the preceding argument they have been treated principally as the results of overrating technical concepts. The scope of stimulus-response behaviorism, or of neodarwinian instinctivism, is much less than some writers (and some readers) have supposed. This is disappointing: it means that we are more ignorant than we should like to be, on difficult questions about ourselves.

Second, evolutionary determinism represents man as driven by ineradicable and disagreeable impulses (or instincts) fixed by natural selection. It exemplifies a misanthropic trend that has been influential for at least two-and-a-half millennia. Such an attitude perhaps often reflects personal feelings of cynicism or depression. Some people, faced with a choice of positions in relation to their fellow human beings, decide for private reasons to adopt one of pessimism. It is easy to see how this can happen among thoughtful people in a period of threat and confusion—even if it is also a time of opportunity.

The third aspect is the social. Here is a remark (slightly shortened) by J. W. Fulbright, then an American senator, quoted in a symposium volume on comparative social behavior (Eisenberg & Dillon, 1971):

> What is important to those of us in the Congress is to feel that we can influence the decisions which affect the future of this country. If we assume that men generally are inherently aggressive in their tendency, it certainly makes a great deal of difference in one's attitude toward current problems. If we are inherently committed by nature to this aggressive tendency to fight, I certainly would not be bothering about all this business of arms limitations or talks with the Russians.

Biological models of man not only reflect existing social and political ideas, but also help to create or at least to confirm them. The extent to which they do this can hardly be measured; but we are justified in assuming that their political influence is not negligible, even if it is only indirect.

One possible consequence of accepting the pessimistic outlook is fatalism: we are pre-programmed by evolution to be acquisitive, aggressive, spiteful, and egocentric; hence we might as well resign ourselves to all the unpleasant consequences. But, while this gloomy picture has been painted by some, others have refused to accept human misery as inevitable: instead, they have taken action against it. We may therefore regard determinist and pessimistic writings, not as descriptions of something preordained, but as warnings. They tell us what to avoid in our own conduct. To paraphrase Marx: The sociobiologists have only interpreted the world; the point is, to change it.

Summary

Current biological models of man are of two types. One, the sociobiological, interprets human conduct by the supposed past action of natural selection. Emphasis is put on disagreeable features such as selfishness, spite, and antisocial violence. These tendencies are treated as if they represented "instincts" unalterably fixed by our evolutionary past. Correspondingly, the term "altruism" (which usually refers to a moral concept), is employed for behavior which aids a conspecific at a cost to the actor. For this notion (which belongs to population genetics) I propose the term *bioaltruism*.

Theories concerning the evolution of social behavior show how natural selection could have produced the known conduct of extant forms. Whether it did so in fact cannot be found out. The theories also provide a valuable framework for the study of existing social patterns in relation to ecological niche. They have few implications for man, if only because the human species has no single habitat or mode.

The second type of model allows that human beings are capable of adapting their behavior to circumstances; but such models are based on simple notions of "conditioning" (Pavlov) or the effects of reward and punishment (Skinner), which are inadequate even as accounts of the achievements of quite lowly animals.

Both kinds of model represent attempts to interpret phenomena of great complexity in terms of simple notions developed by specialists working in narrow fields of study. Both are deterministic. They disregard features that distinguish the human species sharply from all others, and which allow us to make choices. These features include communication by languages which allow argument, abstract concept formation, and fantasy; self-consciousness; and rapid social change based on teaching and tradition.

Pessimistic accounts of the human condition may be regarded as warnings of what we may choose to avoid.

BIBLIOGRAPHY

Allen, G. E. 1975 "Genetics, Eugenics and Class Struggle." *Genetics,* 79:29–45.
Allison, A. C. 1969 "Natural Selection and Population Diversity." *J. Biosoc. Sci.,* Suppl. 1:15–30.
Arieti, S. 1972 "The Will To Be Human." New York: Quadrangle.
Barnett, S. A. 1977 "The Instinct To Teach: Altruism or Aggression?" *Aggressive Behavior,* 3:209–29.
——— 1979 "Cooperation, Conflict, Crowding and Stress: An Essay on Method." *Interdisciplinary Science Reviews,* 4:106–31.

Brown, J. L. 1978 "Avian Communal Breeding Systems." *Ann. Rev. Ecol. Syst.,* 9:123–55.

Chomsky, N. 1959 "Verbal Behavior." *Language,* 35:26–58.

Clark, J. D. 1976 "African Origins of Man the Toolmaker." *In,* G. L. Isaac and E. R. McCown (eds.), *Human Origins.* New York: Benjamin.

Clutton-Brock, T. H., and P. H. Harvey (eds.) 1978 *Readings in Sociobiology.* San Francisco: Freeman.

Dart, R. 1954 "The Predatory Transition from Ape to Man." *International Anthropological and Linguistic Review,* 1:207–8.

Darwin, C. 1873 *The Expression of the Emotions in Man and Animals.* London: Murray.

———— 1901 *The Descent of Man and Selection in Relation to Sex.* London: Murray.

Dickeman, M. 1975 "Demographic Consequences of Infanticide in Man." *Ann. Rev. Ecol. Syst.,* 6:15–37.

Eisenberg, J. F., and W. S. Dillon (eds) 1971 *Man and Beast: Comparative Social Behavior.* Washington: Smithsonian Institution.

Flew, A. 1978 *A Rational Animal and Other Philosophical Essays on the Nature of Man.* Oxford: Clarendon Press.

Gadgil, M. 1972 "Male Dimorphism as a Consequence of Sexual Selection." *Am. Nat.,* 106:574–80.

Gallup, G. G. 1970 "Chimpanzees: Self-recognition." *Science,* N.Y., 167:86–87.

Gardner, R. A., and B. T. Gardner 1975 "Early Signs of Language in Child and Chimpanzee." *Science,* N.Y., 187:752–53.

Haldane, J. B. S. 1959 "Natural Selection." *In,* P. R. Bell (ed.), *Darwin's Biological Work.* Cambridge: University Press.

Hamilton, W. D. 1964 "The Genetical Evolution of Social Behavior." *J. Theoret. Biol.,* 27:1–52.

Jolly, C. J. 1970 "The Seed-Eaters: A New Model of Hominid Differentiation Based on a Baboon Analogy." *Man,* 5:5–26.

Kamin, L. J. 1974 *The Science and Politics of I.Q.* Hillsdale, N.J.: Erlbaum.

Kleiman, D. G. 1977 "Monogamy in Mammals." *Q. Rev. Biol.,* 52: 39–69.

Krebs, J., and R. M. May 1976 "Social Insects and the Evolution of Altruism." *Nature,* Lond., 260:9–10.

Krebs, J. R., and N. B. Davies (eds.) 1978 *Behavioral Ecology: An Evolutionary Approach.* Oxford: Blackwell Scientific.

Lack, D. 1968 *Ecological Adaptations for Breeding in Birds.* London: Methuen.

Lee, R. B., and I. DeVore 1968 *Man the Hunter*. Chicago: Aldine.

Lewontin, R. C. 1979 "Sociobiology as an Adaptationist Program." *Behavl. Sci.*, 24:5–14.

Ligon, J. D., and S. H. Ligon 1978 "Communal Breeding in Green Woodhoopoes as a Case for Reciprocity." *Nature*, Lond., 276:496–98.

Lincoln, G. A. 1972 "The Role of Antlers in the Behavior of Red Deer." *J. Exp. Zool.*, 182:233–50.

Lincoln, G. A., and F. E. Guinness 1973 "The Sexual Significance of the Rut in Red Deer." *J. Reprod. Fert.*, Suppl. 19:475–89.

Lucas, J. R. 1970 *The Freedom of the Will*. Oxford: Clarendon Press.

MacKay, D. M. 1960 "On the Logical Indeterminacy of a Free Choice." *Mind*, 69:31–40.

Maynard Smith, J. 1964 "Group Selection and Kin Selection." *Nature*, Lond., 201:1145–47.

——— 1978 "The Ecology of Sex." *In*, J. R. Krebs and N. B. Davies (eds.), *Behavioural Ecology: An Evolutionary Approach*. Oxford: Blackwell Scientific.

Maynard Smith, J., and M. G. Ridpath 1972 "Wife Sharing in the Tasmanian Native Hen, *Tribonyx mortierii:* A Case of Kin Selection?" *Am. Nat.*, 106:447–52.

Medawar, P. B. 1959 *The Future of Man*. London: Methuen.

Montagu, A. 1976 *The Nature of Human Aggression*. New York: Oxford University Press.

Orians, G. H. 1969 "On the Evolution of Mating Systems in Birds and Mammals." *Am. Nat.*, 103:589–603.

Pavlov, I. P. 1928 *Lectures on Conditioned Reflexes*. London: Lawrence and Wishart.

Peters, R. S. 1974 *Psychology and Ethical Development*. London: Allen and Unwin.

Premack, D. 1976 *Intelligence in Ape and Man*, Hillsdale, N.J.: Erlbaum.

Robertson, D. R. 1972 "Social Control of Sex Reversal in a Coral-Reef Fish." *Science*, N.Y., 177:1007–9.

Rowley, I. 1965 "The Life History of the Superb Blue Wren." *Emu*, 64:251–97.

Russell, B. 1927 *An Outline of Philosophy*. London: Allen and Unwin.

——— 1958 *Portraits from Memory and Other Essays*. London: Allen and Unwin.

Sahlins, M. 1977 *The Use and Abuse of Biology*. London: Tavistock.

Skinner, B. F.   1953   *Science and Human Behavior.* New York: Macmillan.

——— 1957   *Verbal Behavior.* New York: Appleton-Century-Crofts.

——— 1973   *Beyond Freedom and Dignity.* Harmondsworth: Penguin.

Slobodkin, L. B.   1978   "Is History a Consequence of Evolution?" *In,* P. P. G. Bateson and P. H. Klopfer (eds.), *Perspectives in Ethology,* Vol. 3. New York and London: Plenum.

Trivers, R. L.   1971   "The Evolution of Reciprocal Altruism." *Quart. Rev. Biol.,* 46:35–37.

——— 1972   "Parental investment and Sexual Selection." *In, Sexual Selection and the Descent of Man 1871–1971.* Chicago: Aldine.

West Eberhard, M. J.   1975   "The Evolution of Social Behavior by Kin Selection." *Q. Rev. Biol.,* 50:1–33.

Wilson, E. O.   1975   *Sociobiology.* Cambridge: Harvard University Press.

Woolfenden, G. E.   1975   "Florida Scrub Jay Helpers at the Nest." *Auk,* 92:1–15.

Young, R. M.   1971   "Darwin's Metaphor: Does Nature Select?" *Monist,* 55:442–503.

STEVEN ROSE

"IT'S ONLY HUMAN NATURE": THE SOCIOBIOLOGIST'S FAIRYLAND

Football crowds shout for rival teams and, at the end of the match, fights break out between supporters of the two sides. Advanced industrial nations spend up to 10 per cent of their gross national product on armaments of greater and greater sophistication when their stock-piles are already sufficient to obliterate all life on earth many times over. Why? "It's only human nature. Man is by nature aggressive, and these are two ways of showing it."

Schoolchildren compete in exams for top place; adults compete for jobs against a background of unemployment; businesses compete with each other for contracts and profits. Why? "It's only human nature. Man is by nature greedy and competitive and seeks power over others; some people are naturally superior in the competition, others inferior, and the struggle sorts out natural winners from natural losers."

Entry into Britain by foreigners, especially those with black, brown or yellow skins, is restricted by law. Once inside, groups with different religions or skin colours are discriminated against in housing or jobs and their families physically and mentally assaulted. Why? "It's only human nature. Men are by nature territorial, group-living animals, asserting of their rights of ownership over land, and xenophobic."

From *Race and Class,* Vol. 20, No. 3 (1979). Copyright © 1979 by The Institute of Race Relations, London. Reprinted by permission of The Institute of Race Relations.

Throughout society, men occupy the highest positions, in government, industry, science, medicine; women, the inferior ones as secretaries, technicians, schoolteachers, nurses. Why? "It's only human nature. Men have naturally higher skills than women for these demanding tasks—women are essentially nurturative, concerned with child-rearing, and at work only incidentally—and only in the jobs which mimic their roles as home-makers; office wives, tenders of the sick, teachers of the young. In all societies patriarchy is inevitable."

How often, when oppressed people in struggle query some aspect of the social order, does the answer come back like that, full of the heavy certainty of the "naturalness" of any piece of human conduct, that "things are so and rightly and inevitably so." "You can't change human nature," we are told, with an air of either smug satisfaction or pious resignation, when we are moved to protest about any seeming injustice. Yet what is this mysterious, looming abstraction which seemingly lies at the core of any piece of human conduct, any type of social relation?

Up until the middle of the last century, the inevitabilities of human nature were seen as part of god's ordering of the universe, a god which had created humanity in a given mould, provided rules for the proper conduct of human affairs and established an unquestionable hierarchy: "the rich man in his castle, the poor man at his gate." With the final death of god—already wounded fearfully by the rise of Newtonian science in the seventeenth century—at the hands of the triumphant Darwinism of the nineteenth, then science, in the form of biology, replaced him as the arbiter of human nature and destiny. Biology, rather than god, was responsible for setting the limits to human conduct and potential: class, race and sexual struggles were to be seen as the workings out of the inevitable consequences of the iron laws of evolution: the struggle for existence, the survival of the fittest, were the Darwinian categories that lay beneath Victorian laissez-faire capitalism, and Britain's imperial expansion. Social Darwinism, given its full ideological form by Herbert Spencer, was seductive as a mode of describing and rationalizing "the way the world was" not merely to philosophers and politicians, but to many biologists as well, and did not Darwin's own "solution" to the mechanism of evolution derive from Malthus's view of the inevitability of competition for scarce resources in human populations?

Over recent years, after a period of disrepute, this tendency towards biological law-giving has once again become high fashion, dignified now by names which lay claim to new scientific legitimacy. Ethology, which has a long and in many ways distinguished intellectual history, has become aggrandized into "sociobiology," whose prophets claim it as the science of the future into which will be merged as well not merely the brain sciences such as neurobiology and psychology, but also the sciences of human society, sociology, economics, politics. The mainsprings of human conduct will be discovered deep in all of our biological histories and as a result not only will human nature be understood and quantified, it will become predictable as well; from the position you and your partner adopt while copulating and your neighbour's quarrel with his mother-in-law, through the protests over nuclear power to the date and form of the coming revolution in South Africa—sociobiology's claim to the ownership of the inscribed tablets of the iron laws of history are universalistic to the point of megalomania.

It is easy to imagine a better world than the present—one for example, where, across large areas of the globe people do *not* die routinely of famine and famine-induced disease; where the perinatal mortality of Liverpool or the Rhondda is no higher than in Hampstead or Bournemouth; where unemployment and routine alienated labour is eliminated and our children educated to develop their human and creative possibilities to the full; where we live without the imminent threat of destruction from nuclear war begun by accident or design; and where humanity's relationships with nature are harmonious rather than exploitative. Yet at the same time our imagination of, our striving for, the new world runs full tilt into the claims of "hard-nosed realism." What is, is what must be. It is only human nature. Offered a vision of Utopia, the realist defenders of the status quo substitute sociobiology. For them, in the kingdom of the blind, the one-eyed prophets are to be defined as mad, and have their eyes removed.

So it is important to look at the method and reasoning employed by this law-giving subject which claims to tell us who we are and how we must live. In this article, I want to discuss some of the general claims made by sociobiology to be able to provide explanatory or predictive knowledge of the behaviour of humans, individually or in groups.

Now it is relatively easy to respond simply by saying that sociobiology tells us what *is*, but not what *ought* to be: that one cannot derive moral precepts from biological observations; that biology is neutral about human morality whatever it says about human animality—indeed, many sociobiologists, though not all by any means, would offer such a disclaimer. The substance of my disagreement with sociobiology's claims, however, is far deeper than this truism. I would argue that the entire structure of the method and reasoning employed by sociobiology contains a series of fundamental flaws. Hence the general claims that it makes to provide explanatory or predictive knowledge of the behaviour of humans, individually or in groups, are *scientifically* invalid, and not merely morally neutral or otiose. Note that this critique of sociobiology is written from a standpoint which accepts that there is a field of *science* for discussion here—a field which may become, and in the case of sociobiology *has* become a battleground between science and ideology, but one in which, nonetheless it *is* possible to distinguish truth from falsity, science from ideology or social relations.[1] I am specifically *not* discussing here the social determinants of the renewed interest in biological determinism, nor its manifestly ideological functions.

I am setting this discussion against two backgrounds. The first is the surge of popular and semi-popular books which have appeared over the last few years, by people like Desmond Morris, Konrad Lorenz, Robert Ardrey, Tiger and Fox, and more recently Richard Dawkins (*The Selfish Gene*).[2] All of these claim to provide accounts of some of the terrain which E. O. Wilson, in a more academic but no less controversial book, has called *Sociobiology—the New Synthesis* (Harvard University Press, 1975), a pared down and popular version of which is *On Human Nature*. The second background point to be made is that these sociobiologists do not all speak with one voice; indeed there are quite bitter disputes over theory and methodology between them, as between, for example, the school which claims that the "unit of selection" in the evolution of societies is the group, and that which believes it is the individual or the genes packaged inside that individual and his or her near relatives. These disputes do not concern me here. For the purpose of my present discussion, whether one is a group- or kin-selectionist is irrelevant. I want to try to demonstrate that all types of contemporary sociobiological thinking embody within them a set of interlocking fallacies, which render their claim to scientificity spurious. And in order to do this—and this is the final caveat I wish to make—I propose to use examples wider, perhaps, than some advocates of sociobiology would want to accept as falling within their framework. Having exposed these fallacies, I am then able to ask: what ideological function—whose interests—do they serve? Who benefits from sociobiological thinking?

First, what the sociobiologists do is to take aspects of human behaviour and attempt to abstract certain common features from them. A mother protecting her baby from attack, a doctor risking his or her life in a cholera epidemic, a soldier leading a doomed assault on a machine-gun post, all express "altruism." A child's logical and numerical skills, its reading ability and conformity to its teacher's expectations are all measures of an underlying, unitary "intelligence." Note how this trick works, by taking a process with a dynamic and a history of its own involving individuals and their relations, that is a social interaction, and isolating out an abstract, underlying, fixed thing, or "quality." This process, of *reification,* is not dissimilar to the phrenology of the early nineteenth century, in which a human was seen as a mosaic of different pieces of behaviour —a person's brain had a "bump of philoprogenitiveness," another for "love of music" and so forth.

Now sociobiologists would dispute this claim, arguing that far from lumping together disparate activities and reifying them in a phrenological manner, they were doing no more and no less than Newton when he saw the fall of an apple and the motion of the moon as different cases of the same law of gravity. The difference between the sociobiologists and Newton is, however, profound. It is as if Newton had defined gravity as a "property" of the apple, rather than as law describing an aspect of the relationship between the apple and the earth. It was precisely because he transcended the reified property and understood the relationship between objects that Newtonian mechanics advanced, and it is because they do not do this that the sociobiologists' attempt to reduce disparate phenomena to a common denominator must fail. In the sociobiological universe, characteristics such as altruism or aggression, locked inside individuals' heads as reified properties, also become abstract forces which move the individuals who possess them like clockwork mechanisms. One can see this at work when sociobiologists attempt to measure the quality that they have reified, as for example when they attempt to quantify aggression by reducing it to a measure of how fast rats kill mice placed in cages along with them—an experimental approach dignified with the name of the study of "muricidal" behaviour.

What happens in this approach is that the fixed and reified property becomes attached to an individual rather than emerging from a situation. It is individuals who then become aggressive, altruistic, intelligent and so forth, and the *same property* becomes manifest in different circumstances. The aggressive male is essentially playing the same role whether on a picket line, in a football crowd or beating his wife; he is expressing, in different forms, aspects of the same underlying biological property (contrast this, for example, with a definition of "intelligence" not as a property of the individual but of a relationship—of that individual with others and with the social and natural worlds which confront her or him).[3]

Once you've performed this labelling trick, the way is open to seek for the location of the quality inside the individual—for if a person is aggressive, it is clear that there must be, for instance, a region of the brain which is *responsible* for the aggression—from which it follows that one can eliminate "undesirable" aggression, whether of strikers or football fans, by removing or modifying the bit of the brain which is responsible for it.

Examples of this first fallacy, of reification, are numerous. One is the Russian practice of labelling political dissidents as mad and treating them for schizophrenia. A second is a book by a couple of American psychosurgeons. *Violence and the Brain*,[4] which, in the face of the inner city riots which raged through the US, argued that whatever the social reasons which might be involved, the riots might be best explained by there being something wrong with the "aggression centres" in the brains of a number of ghetto leaders, and that urban violence could be cured by removing a small region of the brain—the amygdala—from some 5 to 10 per cent of ghetto

dwellers. A couple of years ago, I came across a copy of a letter
from an American prison governor which made the same case—it
asked for neurosurgical examination and intervention for a black
prisoner whose signs of "illness" were "organizing a prison work
strike, learning karate, hatred for white society and reading revolutionary literature." In similar vein, it is common practice in American inner cities now to define children who are disrespectful of their
teachers and poor learners (with a high proportion of blacks and
Chicanos) as suffering from "minimal brain dysfunction" and to
drug them with an amphetamine-like substance called Ritalin. It
is estimated that some 600,000 pre-puberty inner city school children are currently being dosed with daily Ritalin pills on the basis
of school reports—not because there are any physical or neurological "signs of disorder," but because their teachers and psychia-
trists have reified, out of the complex interaction of these school
kids with their schools, homes and the wider society, a "disease" of
too little amphetamine in the brain!* What this reification does is
to reverse the old slogan "do not adjust your mind, the fault is in
reality."[5] Instead, the "fault" is located within the individual, a
phrenological bump of aggression, or social dissatisfaction.

Having abstracted—reified—aspects of a social interaction into
uniform qualities of an individual participant in that interaction, the
next fallacy of sociobiological thinking is to _quantify the quality_.
That's easy: you simply ask "how much" aggression, intelligence,
altruism or whatever an individual possesses. We are accustomed
to thinking in linear, ordinal terms. If someone is intelligent, we
ask "how intelligent?" "more or less intelligent than some other
person?"—and so on, till we end up with a ranking scale against
which the entire citizenry can be categorized and their "quality,"
intelligence, given a number, like the infamous IQ. The spurious
numerology which masquerades as science has such a grip on day-
to-day thinking that it is often not hard to accept the apparent
scientificity of this type of sociobiological argumentation.

* Although there is some evidence that Ritalin and related substances may be
in very sporadic use in Britain, it is important to emphasize that on the whole
the British response has been to label children as maladjusted or educationally
subnormal and dump them into special schools instead. Therapy seems more
likely to mean attempted behaviour modification by reward and punishment
than drugs.

Asking the question "how much" in an ordinal, linear way, is only very seldom a sensible scientific question and is often quite meaningless—to reply to the question "how many bananas make six" by saying "two apples" may be grammatically correct, but is empty of scientific content.

So there are the first two fallacies of sociobiology—the reification, then the quantification. Now for the third. This is the appeal to biological "evidence" from the study of non-human animals. As humans are a particular animal species, with a continuity of evolutionary links with the rest of the biological world, it is not unreasonable to expect to find analogies for human behaviour amongst other animals. The strength of ethology has lain in its study of animals in their natural environments and in social interaction. However, the problem that such studies run into is just the same as that of the study of human behaviour, but more so, for it is now necessary to look in non-humans for "qualities" abstracted out of human experience. If it is hard enough to study an aspect of human relations without reifying, what happens when we take terms for human behaviour—intelligence, curiosity, altruism, aggression—and look for their analogies in animals? So we find species of ants which introduce others into their nest, where they perform certain of the tasks of the ant economy, classified as "slave-making" ants. Some strains of rats, in appropriate conditions, tend to kill mice placed in their cages, and these become labelled as "aggressive" rats. The use of an identical label, which may be no more than a sort of pun, or observer's shorthand, becomes an explanation. Unitary labels must imply unitary mechanisms, and "aggression" in other animals must imply the "innate aggression" of humans too. Thus a sort of conjuring trick takes place in which the sociobiologist looks at the natural world in terms of human categories, and then turns back to the social world once more and claims that because such phenomena occur in other animals as well as humans, they must be biologically based in humans, and hence they represent inevitable and immutable categories of human behaviour. It is a world of looking-glass logic, to which E. O. Wilson is particularly prone.

The final trick—the fourth of the sociobiological fallacies—is to play a sort of fairy-tale game. In it, one says "just suppose" that any particular human or social character was biologically—genetically

—determined, then what would the consequences be? A good example of this comes in Dawkins's book, in which he uses the "just suppose" game for sexual constancy. "Just suppose," he argues, that all females were sexually genetically constant, whilst a proportion of males had a gene for constancy and the rest a gene for "philandering"—then what would the consequences be? He concludes that a population with a given proportion of male "philanderers" would form an "evolutionary stable strategy," and presumably, therefore, that a male's propensity to be faithful or unfaithful to his spouse is just another of those biologically determined variables of human behaviour which is immune to social explanation. Whether you are (as a male) a Casanova or an Abelard is genetically laid down before you start (so, according to one of Wilson's most favoured acolytes, R. Trivers, may be the alleged like of adults and dislike of children for spinach!). The point is not that one shouldn't make models of behaviour, it is that we are in constant danger of being seduced by our models, of being so enchanted by the fact that they "fit" the data that we ignore that an infinity of *other* models could equally well fit one's observations. The point about models is that they must be testable and refutable. The trouble with the sociobiologist's models is that they become a closed world—there is no sort of situation to which one cannot get a fit granted enough suppositions about genes for this or that piece of behaviour and some other genetic properties like dominance or partial expressivity. Some of the wilder reaches of this sort of fairyland model-building come when we discover that, according to some psychometricians of the Eysenck school, not only are differences in intelligence largely inherited but so are children's capacity to learn French at school, neuroticism, radicalism versus conservatism in political thought and even twins' tendencies to answer consistently or inconsistently on questionnaires (thereby "saving the phenomenon" with a vengeance!).

The charge that the sociobiological world is "closed" to possible experimental refutation, whilst serious, does not make sociobiology unique. Many fields of science operate within paradigms—general overarching theories—which are so comprehensive and adaptable to "new" facts that they are not capable of refutation in the classical sense of philosophy of science—evolutionary theory itself is an example. So is astronomy, whether in its Ptolemnic or Galilean form.

Yet evolutionary theory, despite the unresolved paradoxes within it—and they are many and powerful—is *the* great unifying hypothesis within which so much of biology can be encompassed that its strengths are unchallengeable. My charge against the tautologies and self-fulfilling prophecies of sociobiology is that, because of the set of fallacies embedded within them, which result in the reduction of living organisms in all their richness to jerky caricatured puppets pulled by strings of "selfish genes," its tautologies are scientifically sterile*—the tautologies of pre-Copernican astronomy, not those of evolutionary theory. Far from being much abused "new Galileos," as their advocates have claimed for them, the sociobiologists are mere Ptolemaic medieval schoolmen. Finally, sociobiology's claim to biologize away the study of human societies by the methods of sociology, economics, political science, and history, fails the very simple test that there is no way that sociobiology can possibly account for either differences between particular existing human societies—for instance between South Africa and Tanzania—nor rapid changes of social form within a given society—for instance between China pre-1948 and after 1968. If it can't do these things, at best the exercise becomes a piece of fashionable Harvard or Oxford intellectual games-playing; at worst a way of ideologically justifying the status quo. These reflections on the methodology of sociobiology in general can be seen as applying to Wilson's *On Human Nature* in particular, as the book represents a pared-down version of his earlier *Sociobiology,* emphasizing how the "new" thinking of sociobiology affects our understanding of human societies, though a substantial proportion of the book has appeared earlier in magazine and journal articles and a part of it merely reprints a section of the final chapter of *Sociobiology.* The latter book was heavily criticized both for its ideology and its biological and anthropological claims.[6] Nowhere in the present book does Wilson acknowledge these criticisms explicitly and nowhere in his quite extensive bibliography does he refer to a single one of the many articles and books published by his critics. Despite this sur-

* Note that, as I emphasized earlier, I am saying nothing here about the ideological function of these sociobiological propositions, but rather viewing them from the "inside" of a science-versus-pseudoscience battleground.

prising lack—of academic courtesy if nothing else—Wilson's claims in the present book are moderated from some of the more incautious phrases of *Sociobiology*. Gone, for example, are "genes" for "spite" or "indoctrinability" or whatever, in the crudely phrenological style of his earlier book; now caveats surround each assertion to the point where it often virtually disappears into a cloud of "tendencies."

The book takes us through a series of topics which are the now familiar hunting ground for sociobiological theorizing and which have been discussed above; the evolution of human society, aggression, sexual behaviour and differentiation, "altruism," and religion. It concludes with a call for hope, based upon what he describes as the "seemingly fatal deterioration of the myths of traditional religion and of its secular equivalents," chief amongst which he places Marxism, and their replacement by an ethic of the search for "scientific truth," previously called for in *Chance and Necessity* by the molecular biologist Jacques Monod[7]—the sort of philosophical naïveté that the French immediately dubbed Monodtheism. In Wilson's case this call is the less credible in that whilst he has clearly read his Bible, he shows little evidence of even having read *Marx for Beginners*.

What Wilson wants to claim is that the organism and its behaviour represent no more than the gene's way of making another gene; that is, the richness of our biological, mental and social life may be reinterpreted simply in terms of strategies for the survival of selfish genes. This is what we may call hard sociobiology. The fallacy inherent in this neat formulation (deriving, I believe, from Samuel Butler, who argued a century or so ago that a chicken was merely the egg's way of making another egg) is seen if the paradox is inverted; after all it is equally plausible, paradoxical, and fallacious to argue that the gene is merely human behaviour's way of creating another piece of human behaviour.

In the book, Wilson slips uneasily between defence of this hard sociobiology and a much softer version. Sometimes he does this explicitly, as when he claims there are two forms of altruism, hardcore and softcore, and sometimes implicitly, when he lowers his sights to something little more contentious than the claim that one cannot understand human society in the absence of an understanding of human biology. Now while it is easy to understand the relevance of such a soft claim in reaction to the sort of sociological or psychological reductionism (for instance Skinnerian behaviourism) which has been an important intellectual strand in the US, or even in response to some of those ideologues who wish to debiologise the human condition entirely,[8] it is scarcely a serious theoretical challenge to any sort of Marxist thinking.

The real failure of the sociobiologists lies in their seeming inability to avoid the either/or trap. Behaviour must be *either* socially *or* biologically determined, or must represent the arithmetic sum of a biological (genetic) and an environmental component. On the contrary, a proper understanding of the interaction of the biological and the social in the production of humans and their society will only be possible following the simple recognition that *both* genes *and* environment are perfectly necessary to the expression of any behaviour. That humans have two legs and speak depends upon their genotype; if they were, say four-legged and incapable of spoken communication, human society would be very different. Hence human society is genetically determined. This proposition, which is all soft sociobiology boils down to, is trivial. But intermingled with the soft sociobiology are the harder Wilsonian claims.[9]

Perhaps most of human sex differences in cognition are environmentally determined, but an itsy-bitsy is genetic and in favour of the men, and if "we" "choose" a society which minimizes these differences "we" may; but in doing so "we" will be going against nature (note Wilson's deliberate avoidance of the question of who "we" are in this context). In the same way, although he does not go as far as Dawkins in proposing sex-linked genes for "philandering," for Wilson human males have a genetic tendency towards polygyny, females towards constancy; having a bit on the side is a male characteristic, while females who are sexually attractive have a genetic tendency to rise upward through the social classes. Genetic determinism constantly creeps in at the back door.

The aim of a truly human science of sociobiology would not be to debiologize or to desocialize our understanding of the human condition. It is vital that an integrated dialectical account of human nature be achieved and the traps of either biological or sociological reductionism avoided. Wilsonian sociobiology is, however, a travesty of this goal, marred by grandiloquent claims, falsely dichotomous thinking, and an incapacity to distance itself from the particularist assumptions of the dominant racist, sexist, and class bound ideology of late twentieth-century western society.

It is against this that we must pose a real science and vision of humanity—one which says that it is the biological and social nature of humanity to transform itself, reach beyond itself constantly: that what seems fixed or constant is so only in the historical moment which itself is always in flux, that the human nature of feudal, preindustrial society was not the human nature of the industrial revolution, is not the human nature of today's advanced capitalism —and will not be the human nature of the transformed societies of tomorrow—those that will at length have truly achieved that old goal, the freedom of necessity.

NOTES

1. On this discussion see Hilary Rose and Steven Rose, "The Metaphor Goes into Orbit: Science Is Not All Social Relations." *Science Bulletin* 21 (1979).
2. *The Selfish Gene.* London, 1976.
3. See Hilary Rose and Steven Rose, "The IQ Myth." *Race & Class* (July 1978).
4. V. Mark and L. Ervin, *Violence and the Brain.* London, 1970.
5. For more details see the discussion in Hilary Rose and Steven Rose, *The Political Economy of Science.* London, 1976.
6. For instance, the various publications of the US Science for the People Study Group on Sociobiology, including their review in the *New York Review of Books* and, most recently their book, *Biology as a Social Weapon.* Cambridge, Mass., 1978, or the very useful anthropological demolition by Mashall Sahlins, *The Use and Abuse of Biology.* London, 1977.
7. *Chance and Necessity.* London, 1971.
8. See "The Metaphor Goes into Orbit."
9. See my review of *On Human Nature* in *New Scientist* (October 19, 1978).

JEROME H. BARKOW

SOCIOBIOLOGY: IS THIS THE NEW THEORY OF HUMAN NATURE?

Is sociobiology a theory of human nature and society? Does it surpass the systems of Marx and Freud in breadth and insight while retaining the genuinely scientific base these older approaches merely claimed? The controversy over sociobiology has arisen over its enthusiastic application to human nature and society, after all, not over its adequacy in explaining langur infanticide or the evolution of alarm calls. This article will explore the applicability of sociobiology to our own species. It will conclude that the approach does not now constitute a scientific theory of human behavior and society but that it does raise the exciting possibility of the vertical integration of the social-behavioral sciences on a foundation of biological evolution.

What Is Sociobiology?

Though Wilson originally proposed the term as a grand synthesis of biological-behavioral fields united in the common framework of evolutionary biology, "sociobiology" has come to mean the distinctive approach taken by a small group of evolutionary biologists.[1-13] These theorists have explored the implications for behavior of certain recent corrections in evolutionary theory.

One now-corrected error involved the level at which selection occurs. Williams and others have shown that evolution rarely takes place in terms of the differential survival of entire groups competing with one another.[14] Therefore, explanations of individual behavior in terms of its adaptiveness for the group as a whole tend to be invalid. Such behaviors may indeed have value for the group as a whole, but that is seldom why they were selected. Thus, Sherman[15] has found for Belding's ground squirrel that while alarm calls tend to aid the group as a whole, they evolved not for this reason but because they help the giver of the call to maximize his genetic representation in the gene pool, that is, his *inclusive fitness.*

This last concept, developed most fully by Hamilton, is the key to sociobiology.[4, 5] Individuals share genes with their kin, the exact proportion depending on the extent of inbreeding but, for example, never dropping below ½ for offspring or (on the average) for siblings, ⅛ for cousins, and so on. Evolution selects for behaviors that result in the maximization of the genes underlying them, regardless of the fate of the particular individual carrying these genes. I may represent 100 per cent of my own genes but my 13 cousins together total 104 per cent of these same genes: selection will therefore favor my lowering my personal (Darwinian) fitness to protect this group of cousins, since such action is likely to result in a net increase in the frequency of the genes underlying it within the gene pool as a whole, even if I am killed. As a product of this *kin selection,* I (and other organisms) tend to act to maximize our genetic representation in the gene pool, rather than our personal survival. The model used is of the cost-benefit kind familiar to economists.[16] Careless reading of sociobiology can give the impression that the major "instinct" of organisms is mathematics and that we all go about computing coefficients of consanguinity (proportion of shared genes) in calculating the relative advantage of investing in one individual rather than in another. Sociobiology, of course, tells us only that organisms act *as if* they had performed such calculations, since the relative frequency of the genes underlying their behavior has indeed been determined by just such mathematics working itself out over the course of evolutionary time.

The deceptively simple premise of sociobiology—that individuals strive to maximize their inclusive fitness—has proven enormously powerful in explaining the evolution of animal social behavior. Since human beings are also animals, it follows that sociobiology should be equally powerful in accounting for our behavior. The entomologists and other biologists who have developed sociobiology have not resisted the temptation to apply their framework to *Homo sapiens,* thereby generating strong emotions in many social and behavioral scientists. Some of the latter have labelled sociobiology "biological reductionism" and "genetic determinism" the better to dismiss it. But sociobiology cannot be so easily ignored. The field either has accounted or can readily account for the evolution of such human phenomena as altruism, parent-offspring conflict, the double standard, lying, sex differences in behavior, ethnocentrism and race prejudice, incest taboos, altruism, sexual jealousy, and the tendency, in many societies, for a man to pay more attention to his sisters' children than to his own.[1-13, 17] Thus, sociobiology begins to resemble a theory of human nature. I will briefly summarize how sociobiologists account for some of these phenomena before going on to criticize such explanations and discuss what a more complete theory of human nature and society would look like.

How Sociobiologists Account for Human Behavior

Altruism is a central topic in sociobiology and a fitting place to begin. Hamilton explained its evolution in terms of nepotism or kin selection.[4-5] To the extent that we share genes with others, selection will favor our aiding their survival and reproductive success. Earlier approaches to evolutionary theory readily explained why we should be selfishly competitive, but Hamilton makes biological sense of helpfulness toward others, provided they are kin. What of altruism toward non-relatives? Trivers deals with this broader level of helpfulness by considering it as "reciprocal altruism."[6] We should be willing to aid non-kin *provided* they are likely to reciprocate our aid either to ourselves or our kin. Thus, reciprocal altruism, found chiefly among human beings, is merely another way for us to maximize our own inclusive fitness. The theory of reciprocal altruism is less elegant and more complex than that of nepotistic altruism, since it must account for selection against non-reciprocators ("cheaters") and explain how the behavior could initially evolve. Nevertheless, the explanation of the evolution of reciprocal altruism is a key triumph of sociobiology.

Sociobiology deals more straightforwardly with ethnocentrism and race prejudice. In effect, the theory is the converse of altruism. We should learn readily to dislike and withhold cooperation from those with whom we do not share genes, especially if we do not have the sort of long-term relationship with them conducive to the development of reciprocal altruism. Among human beings, in the absence of genealogical knowledge, the best indicator of consanguinity is physical appearance. We are more likely, all things being equal, to share genes with those who look like us; and we are more likely to be able to count on our altruism being reciprocated by those with whom we have established long-term ties. Thus, we should most readily learn distrust and hostility toward those who least resemble us, and toward those with whom we have no personal relationship, that is, strangers. It seems clear from the literature on ethnocentrism and race prejudice that we do indeed find it very easy to learn to dislike those who differ from our "ingroup" appearance, or with whom we are unfamiliar.[18] Just as we would expect from the theory, the best counter to prejudice is the establishment of long-term, personal ties to members of the outgroup.

On the other hand, a frequent human reaction to the stranger, even the physically distinctive stranger, is to find him or her attractive—prejudice generally develops only when the strangers are present in some numbers. Here, too, sociobiology can cope: many strangers are likely to compete for scarce resources, so selection should favor non-cooperation or even hostility toward them. A small number of strangers, however, can provide the advantages of outbreeding (heterosis). Presumably the stranger, coming from a different gene pool, is less likely than an ingroup member to share our deleterious recessives. Hence, we find isolated strangers, particularly of the opposite sex, not only non-threatening but even sexually attractive; while we are more likely to be hostile toward same-sex strangers, especially when they arrive in large numbers.

This argument is the converse of Bischof's analysis of the incest taboo.[19] Though cultural systems pattern the incest taboo in many ways, Bischof's review of this topic concludes that we and other species find it relatively difficult to establish sexual bonds with those with whom we have been socialized. In evolutionary time, he argues, such individuals are likely to have been those with whom we shared a high proportion of our genes, so that they also tended to share our deleterious alleles. Thus, those for whom familiarity bred sexual contact tended to eliminate their genes from the pool by having less fit children. Our own ancestors in part became so because they preferred to select their partners from among non-kin and even strangers.

Sexual behavior in general is so major an aspect of human behavior, and sociobiology has so much to say about it, that it can only be touched on here. Robert Trivers's "parental investment theory" has profound implications for sexual dimorphism in behavior.[7] The theory states that the sex which invests more of its total reproductive potential in a single fertilization will be the most discriminating in selecting a partner, an idea which follows logically from the maximization-of-inclusive-fitness premise of all sociobiology. For most species, including our own, it is the female which risks more of her reproductive potential in a fertilization. An ovum costs more, physiologically, than does a spermatozoon. Still more expensive is fertilization: the females' ability to reproduce will be tied up for the nine-month gestation period plus (usually) for the 1 to 3 years before the infant is weaned. The male's investment in a fertilization, in contrast, is almost zero. Thus, selection will favor the more discriminating females and the more promiscuous males. The interests of the two sexes are clearly quite different, a fact that results in courtship.[20]

As Trivers and Wilson make clear, courtship is largely about the selection of males by females.[7, 11] As Chagnon finds for the Yȧnomamö, reproductive success is likely to be far more variable among males than among females, since most of the latter will at least be able to find some partner.[21] Males must therefore compete among themselves far more than must females. Selection favors those females who choose a mate who bears the "best" genes and who is ready to match at least some of the female's greater investment by helping to provide for the eventual children. Thus, male competition—courtship—consists of displays of physical prowess and healthy appearance intended, from an evolutionary point of view, to convince females of the superiority of the underlying genotype. Male competition also consists of amassing the political and economic power which is a good guarantee of ability to invest in off-spring. Finally, males attempt to convince females of their willing-ness to so invest by claiming to be already bonded to them (tradi-tionally taking the form, in our own society, of protestations of love and insincere offers of marriage). Sociobiology predicts that there will be less rivalry among women than among men, and that selec-tion will favor females better able to discern the accuracy of male claims to be already bonded to them, as well as to discern the re-liability of displays of physical condition and economic-political success.

This argument can be taken even further. Trivers argues that the point at which the male is most likely to leave is the point at which he has invested least in a fertilization: immediately after copulation.[7] At this instant, however, the fertilized female's reproductive poten-tial for the next several years is in effect being held hostage. Thus, selection should favor males who wish to decamp or at least change partners following ejaculation, but females who at that point en-deavor to strengthen the male's bond to them. Females should therefore demand tenderness and show of affection, at this time, and males should exeprience a feeling which the Romans diplo-matically referred to as "sadness" (*animal post coïtum triste est*).

That the male post-ejaculatory refractory period is appreciably shortened if a new female is introduced supports these interpreta-tions.

Sexual jealousy, too, is interpretable in terms of sociobiology. Females always have an advantage in being absolutely certain of the identity of their offspring, while for males, paternity certainty is problematic.[1, 12, 17] Males alone run the evolutionary risk of in-vesting parentally in someone else's offspring. Thus, selection should favor any mechanisms likely to increase paternity certainty, and jealousy would appear to be such a mechanism. The least jealous males would have been likely to have had fewer offspring than other males. (Of course, the most jealous males may have suffered a similar fate by driving away potential mates, so we need not ex-pect selection to have favored insane male jealousy, either.) Males should be most jealous over a partner's copulating with another, while females should be most jealous not so much over a partner's copulating as over his investing in another female's offspring (since the first female loses little, in evolutionary terms, if the male limits his investment in others merely to copulating with them).

When average paternity certainty is very low, in a particular society, the average male may be more closely related to his sisters' children than to those ostensibly his own.[1, 12, 17] In such situations, males should divide their "parental" investment between their "own" children and those of their sisters. Kurland and Alexander point out that, in accordance with prediction, matrilineal societies both place relatively little emphasis on female fidelity and chastity and encourage males to invest more heavily in sisters' children than in the wife's. Thus, the mother's brother-sister's son relationship (the avunculate) appears to be consistent with sociobiological prediction.

Moving away from sexual behavior, Trivers's parental investment theory also provides an elegant explanation of weaning conflict.[7] At a certain point in development, the mother can maximize her inclusive fitness by weaning the current offspring in order to hasten a new fertilization. The infant, after all, has only (a minimum of) 50 per cent of the mother's genes, and the new infant will also carry (a minimum of) 50 per cent. But the infant at the breast has 100 per cent of its own genes. Thus, mother is risking 50% of her genes by weaning early in order to produce another 50% of these genes, but she is risking 100% of the nursing infant's genes. Selection will therefore favor infants who try to prolong nursing and mothers who try to shorten it. The risk weaning poses to the infant's 100% of its genes will be reduced with time, of course, until it is outweighed by the increase in inclusive fitness the birth of a sibling would bring. At that point, the infant should wean itself. Thus, mothers should try to wean earlier than infants themselves prefer but, given sufficient time, the infants will self-wean.

Finally, why do human beings lie to one another? So long as evolutionists thought in terms of group selection, this question was difficult—society would be far more efficient if we could speak truth only. Since we share only a proportion of our genes with one another, however, maximization of inclusive fitness implies that selection favors our saying that which will increase our genetic representation in the next generation, rather than that which is true. This means that selection favors the *capacity* to tell the truth but not the imperative. Alexander goes so far as to suggest it may be that "human society is a network of lies and deception."[3]

Invalid Criticisms

The previous section illustrated the claim that sociobiology can appear to provide a theory of human nature and society. The analyses provided are open to criticisms, however, which will be the subject of the following section. But sociobiological explanations are *not* open to certain other criticisms.

For example, it is invalid to argue that modern contraception and other changed circumstances of contemporary life have made sociobiological arguments irrelevant. Evolution always involves adaptation to past and not present environments. We are now in an environment vastly different from the one in which we evolved and most of the behaviors discussed above are quite likely no longer maximizing anyone's fitness. But have we been out of our evolutionary environment for so long that the genetic substrate of human behavior has drastically altered? This seems unlikely, though ongoing evolution may well make such changes in us in the far future. Thus, to point out that sexual jealousy is unlikely to maximize inclusive fitness today is as meaningless for its sociobiological interpretation as it would be to challenge the idea that a lion is a hunter because lions in zoos do not hunt. Of course, if lions live in zoos for a sufficient number of generations they may no longer be hunters; and some day, we may no longer be capable of sexual jealousy.

It is also invalid to argue that sociobiology is dependent upon a rigid genetic determination of behavior, while human behavior is so broad and flexible that it is impossible to speak of a genetic substrate for any particular behavioral category.[1, 11] If human behavior lacks a genetic substrate then it did not evolve, and we have passed from biology to Creationism. Evolutionary biology does not require that there be a direct and inevitable link between a particular gene and a particular behavior, even if careless reading (and writing) of sociobiology may occasionally permit that impression. Evolution is a probabilistic process and will take place if the behavior being selected for has its probability-of-occurrence affected in any way by a genetic substrate in a given environment. In any particular case, the ontogenetic manifestation of the behavior is likely to be highly dependent on specific features of the environment and their inter-action with the genotype. Thus, evolved behavior is probabilistic rather than deterministic. In our own species, evolution has selected for enormous flexibility in behavior. Much of this flexibility is mediated by learning preferences. When we find the taste of sugar agreeable and that of vinegar unpleasant, we are responding to preferences that evolved because they once led to behaviors that enhanced the inclusive fitness of our ancestors. This kind of learning preference is clearly not a rigid instinct.[2] To argue that we learn to dislike those unfamiliar in appearance more readily than we learn to dislike those who resemble us more closely differs sharply from a contention that racial prejudice is "wired in" and unresponsive to experience and learning.

Finally, it is invalid to accuse sociobiology of being a contemporary form of social Darwinism and racism. Absolutely nothing in sociobiology supports the contention that some groups are "biologically superior" or "inferior" to others and that their present culture is a direct reflection of allele frequencies in their gene pool. Sociobiology does admit the possibility that small, genetically isolated groups with distinctive cultures can differ psychobiologically in minor respects from other populations, a hypothesis stemming from the genetic assimilation of culturally acquired characteristics theory. This suggestion, which I discuss elsewhere at some length, follows logically from the belief that human behavioral characteristics have been generated by biological evolution and are not products of supernatural intervention.[22, 23] It has nothing to do with "superiority-inferiority."

Valid Criticisms

Valid criticisms of the easy application of sociobiology to our own species stem from the fact that: 1) the approach is limited to selective pressures and does not predict the mechanisms through which these pressures are mediated; 2) the approach is at times applied without sufficient consideration of ecologically generated selective pressures which, in evolutionary time, may have drastically altered the patterns otherwise predicted by sociobiology; 3) the total combination of selective pressures has produced an information-processing capacity in *Homo sapiens* which has emergent properties not reducible to any theory of selective pressures; 4) this individual-level capacity has permitted social interaction, over time, to generate supra-individual (not "super-organic") systems called "cultures" which themselves evolve in historical processes that, while perhaps lawful, are not reducible to "nothing but" considerations of inclusive fitness.[24]

Any given human behavioral trait can and should be explained simultaneously at several levels of organization. Each of these explanations must be compatible with one another but definitely not identical.[24] For example, weaning and weaning conflict may be interpreted sociobiologically, as we have seen, as products of the different interests of mother and child. The point at which weaning would maximize the mother's inclusive fitness comes earlier than the point at which it would maximize that of the child. Thus, infants protest weaning at a time when the mother is quite prepared to transfer him to other foods. Let us call this explanation "evolutionary." Wilson would classify it as an "ultimate" explanation, and contrast it with a "proximate" equivalent.[11] This last would involve the physiological-hormonal mechanisms that control lactation in the mother and biting-teething in the child. For Wilson, these two levels of explanation would be sufficient. The proximate level would essentially consist of the mechanisms mediating the selective pressures predicted by the ultimate level, and both explanations would be distinct but compatible.

But two levels are not enough. The way a particular woman reacts to lactation and weaning is a product of her individual learning. The infant's response is at least in part a product of his personal experience—has he been fed on demand or on a schedule, has he had supplemental feedings of other foods, what is his nutritional and health state, and so forth. Thus, we can explain weaning behavior at a third level of organization, that of individual experience. This level can include, if one wishes, personality theory interpretations of the causes of the mother's weaning behavior and the later effects of the process on the child's future behavioral development.

Three levels of explanation of weaning are still inadequate. Some societies encourage early weaning and some late. Some wean the child by applying harsh substances to the nipples or sending a child to live with relatives, and others do not. Some societies force-feed almost from birth and others start supplemental feeding much later. In order to understand how cultures pattern and organize feeding and weaning it is necessary to have a theory of sociocultural evolution. Much of the variation in weaning behavior can be related to particular ecological factors, e.g., the kinds of labor women perform, the presence-absence of a post-partum sex taboo (which is in turn related to polygyny, which is in turn related to climate and economy), and so forth.[25] So we need a sociocultural-level theory of weaning.

Almost any item of human social behavior can be explained in at least four different ways: in terms of evolution, physiological mechanisms, individual experience and psychology, and cultural organization. All of these ways must be compatible with one another, but none is likely to be predictable from another. A theory of human nature and society would require this kind of vertical integration of levels of analysis.

Unfortunately, academic disciplinary chauvinism tends to make reductionists of us all. Reductionism consists of claiming that an explanation at a single level of analysis is entirely adequate and the other levels meaningless.[24] The warning flag for the presence of reductionism is the implied phrase, "nothing but"—e.g., weaning is nothing but a matter of individual psychology, or hormones, or a conflict of interest between mother and child over the maximization of their respective inclusive fitnesses. Much of the resistance of social scientists to sociobiology stems from their traditional sociocultural reductionism, that is, the belief that only explanations at the sociocultural level are meaningful. On the other hand, there is little evidence that the sociobiologists have appreciated the need for cultural-level explanations compatible with but not reducible to evolutionary biology.

Sociobiology is inadequate as a theory of human nature and society because it is limited to a single level of explanation, the evolutionary.[24, 26] The sociobiologists have generally ignored other levels of explanation as not germane to their work. Thus, the human ethologist Blurton Jones laments that sociobiology tells us nothing about the mechanisms mediating the selective pressures predicted by evolutionary theory.[27] The ethological literature makes it clear that these mechanisms are likely to be both subtle and intricate, dependent at each step on interaction with particular aspects of the environment.[28]

Yet even at the level of evolutionary explanation, sociobiology is inadequate as a theory of human nature. A species is shaped by the interaction of selective pressures emanating from two sources: competition among individuals each of whom strives to maximize his inclusive fitness; and the advantages of more efficient adaptation to the environment. The effects of the two kinds of selection may at times oppose each other, and the resulting evolution will always be a sort of resultant of forces.[24] But sociobiologists have paid scant attention to the kind of environment in which our species evolved. They are not paleontologists and, in any event, there is still much we do not know about that environment.[29, 30] Perhaps this is why sociobiology has had so little to tell us about the evolution of our enormous flexibility of behavior, capacity for language, or symbolic ability. Applying sociobiological theory in this environmental vacuum may be leading us into serious error. Let us take the issue of male promiscuity and female jealousy as an example of how different pictures of early hominid ecology lead to different ideas about human nature.

We have seen that sociobiology predicts that males should be relatively promiscuous and females should attempt to bond carefully selected males to them. In some species, however, ecological factors make it unlikely that the young will survive without parental investment from *both* sexes. In these cases the sociobiologists find that male investment in the young may equal or even exceed that of the female.[7, 11] Our own ancestors clearly evolved on the savanna, and male protection for females and young was presumably as important for defense from predators as it is among modern savanna baboons. The question is, were the males as a whole protecting the females and young as a whole, and providing them with meat from the hunt in similar fashion? Or were defense and meat-provision on a one-to-one basis, with mated pairs or one-male groups being the basis of social organization (as with the contemporary Hamadryas baboon)?

If our early ecology made it possible and adaptive for a single male and one or several females to exist as a unit apart from the troop, then we would expect selection to have favored strong pair bonds between males and females, and equally strong bonds between males and the offspring of their mates. On the other hand, if defense and hunting were possible only by groups of cooperating males, then bonds and jealousies between the sexes should be relatively weak and males should have no strong "paternal" feelings toward any particular infant but only a generalized feeling of protectiveness toward all young.

I will make no effort to resolve this issue of whether our ancestors more closely resembled a savanna or a Hamadryas baboon. It could well have been that we utilized both sorts of adaptations serially during our phylogenesis, or were selected for the ability to move from one to another in accordance with varying conditions. The point remains, however, that our conception of human nature depends upon our reconstruction of hominid behavioral evolution. So far, the sociobiologists have made no effort to deal systematically with the relevant literature. Thus, the sociobiological predictions about human nature provided earlier may or may not be accurate, but they are certainly premature.

One final example to illustrate sociobiology's present weakness as an evolutionary theory of human nature. Kin selection and reciprocal altruism "explain" why we should be kind to one another. But if we were evolving as cooperative hunter-gatherers, then selection in favor of altruism was much, much stronger than one would otherwise expect. Given that small human groups are likely to be heavily inbred, kin selection would have eliminated tendencies for individuals to be reluctant to join in mutual defense and hunting. Wolves, for example, having also evolved as cooperative hunters, are altruistic to the point of sharing food with one another (by regurgitation feeding).[31] If we evolved under this kind of ecology, the sociobiologist's suggestion that human society may be "built on lies" is a gross exaggeration.

Emergent Phenomena

At the levels of both the individual and his culture, human behavior has emergent properties not reducible to any theory of biological evolution: These properties are generated by elements which in themselves are indeed products of natural selection. There is nothing mystical or paradoxical in the preceding statement. An emergent is a property of a system not reducible to or necessarily predictable from its component elements.[32] Both chemical compounds and human societies have characteristics that are not merely the sum of the traits of their constituent elements. Sociobiologists have paid scant attention to emergents because of their lack of concern with the selective pressures generated by early human ecological adaptation, their slighting of the complexity of the proximate mechanisms mediating selective pressures, and their disregard of the process of sociocultural evolution.

Hominids evolved in a host of complexly interacting and poorly understood selective pressures which are often discussed collectively as selection for "cultural capacity." This capacity has emergent properties. Adaptation to group defense and predation on the savanna apparently resulted in selection for elaborate information-processing abilities which may be conceptualized as internal representations or cognitive maps of the external environment. Adaptation to the increased social dependency of savanna life would have brought equal pressure for the mapping of complex social relationships. The resulting map would necessarily have included an internal representation of the organism itself, giving the system the emergent property of self-awareness.[26, 29, 33] (This "self" would not be a homunculous, as Dawkins fears; presumably, it would include only aspects of the organism that affected fitness).[13]

This article is obviously not the place to develop a theory of human cognition and self-awareness which, in any event, I discuss at greater length elsewhere.[26, 29, 33] Nor is everyone ready to accept the utility of conceptualizations of human information processing in terms of cognitive maps and a sense of self. My point is simply that fairly sophisticated theories of human personality—even those using "soft" and emergent concepts such as "self"—can be quite compatible with sociobiology. In point of fact, without them sociobiology is quite incapable of dealing with the complexity and richness of human behavior. Art, literature, and non-reciprocal–non-nepotistic altruism are products of emergent characteristics of human psychology quite *compatible* with evolutionary theory but definitely not *predictable* from it. We cannot reduce all of human behavior to considerations of inclusive fitness.

Culture is the locus of the second set of emergent properties slighted by the sociobiologists. Any text on cultural ecology will convince the reader that cultures, too, evolve, and in a fairly orderly and even predictable manner.[34] A given technology or subsistence economy or climate has complex ramifications for forms of social organization and even for major cultural values. Thus, Edgerton, in a carefully controlled comparison of East African pastoralists and farmers, found that the farmers were more fatalistic and stressed obedience training for their children, while the pastoralists emphasized personal independence, decisiveness, and self-reliance.[35] These differences, which held true both within each of the four societies and also across them, are readily explained in terms of the behaviors required to make a living in the particular economies in question. Trying to account for Edgerton's findings in terms of evolutionary biology would resemble explaining a problem in organic chemistry purely in terms of particle physics. Edgerton's analysis appears entirely adequate and is compatible with sociobiology without being reducible to it. His work illustrates the principle that laws of sociocultural evolution are emergent phenomena.

Sociocultural phenomena are never fully explained on the sociobiological level, for cultural evolution has emergent properties of its own. Culture *is* the means by which *Homo sapiens* adapts to its environment, but it is never *nothing but* a means of adaptation. Even when culturally ordered behavior is consistent with the maximization of inclusive fitness premise of sociobiology, explanations in terms of cultural evolution (or history) are still needed: How else do we account for the cross-cultural variability of behavior?

Toward a Human Sociobiology

Wilson's original conception of sociobiology was not limited to the implications of recent developments in evolutionary theory: he was trying to synthesize a *discipline* incorporating all biological and sociological approaches to the understanding of vertebrate and invertebrate behavior. His treatment of human beings, compared with his discussions of other species, is almost incidental. Perhaps it should have been omitted entirely: as we have seen, a sociobiology of *Homo sapiens* remains to be written.

A human sociobiology would be not *a* theory of human nature but a synthetic discipline vertically integrating biological, sociological-anthropological, and psychological levels of explanation, much as I sketchily did earlier for weaning. The field would begin with the same principles of evolutionary biology Wilson demonstrates are so effective for understanding other species, and would retain his comparative, cross-species approach. It would pay particular attention to our nearest relatives, the non-human primates, and also to social hunters such as the canids, whose evolution may parallel our own in certain respects. Because most human beings today live in an environment very different from that in which we evolved, a human sociobiology would concern itself with reconstructions of earlier ecologies, seeking to understand the kinds of selective pressures to which our ancestors were subject. In this, the ethnographic study of extant hunting-gathering peoples would be helpful. But our lack of definitive knowledge of protohuman ecology and social organization would always be a weakness in a human sociobiology (though by no means an insurmountable one).

That discipline would incorporate the three remaining, non-evolutionary levels of explanation, just as Wilson's conception of a sociobiology incorporates the other biological disciplines.[36] Human ethologists and developmental-physiological psychologists would be challenged to demonstrate the compatibility of their theories with our (limited) understanding of human phylogenesis, and would undoubtedly find themselves generating new hypotheses in the process. The evolutionists, at the same time, would be obliged to show that their reconstructions were compatible with the findings of the ethologists-physiologists. Such mutual accommodation is circular only in appearance: in practice, standard empirical hypothesis testing for the ethologists-physiologists on the one hand, and the method of comparative analysis for the evolutionists on the other, would serve to keep our logic honest. The personality theorists would be faced with an additional challenge, for they would be obliged to demonstrate that their theories were compatible not only with the reconstructions of the evolutionists but also with the findings of the ethologists-physiologists. On the other hand, the concept of emergence would save them from the "nothing but" reductionism which so often alienates our confreres in the humanities.

The sociologists and social/cultural anthropologists would face the severest challenge, for these disciplines have long suffered from a historic loathing of explanations at the biological and psychological levels. Now they would be expected to demonstrate that their theories were compatible with the conclusions of evolutionists, physiologists, and developmental psychologists, and even personality theorists! Ideally, a social psychology would develop demonstrating how individuals, bearing behavioral predispositions and learning preferences produced by biological evolution, interact with each other in such a manner as to produce and maintain social-cultural forms.[37] While this kind of processual, generative approach already exists in anthropology, usually associated with the work of Fredrik Barth and his followers, its psychological-level of analysis is generally intuitive, implicit, and unformulated.[38]

If each level of analysis of a human sociobiology would be heuristically distinct, actual theorizing would move across levels. For example, there is now a small but growing body of literature exploring the extent to which culturally patterned behavior maximizes the inclusive fitness of its participants. One such study suggests that being successful by the standards of one's own culture is associated with biological success in terms of inclusive fitness—the wealthy and prestigious often have more children than do others, or use their power to aid kinsmen.[39] How would a human sociobiology interpret such a finding?

The answer would have to be that learning preferences, as shaped by biological evolution and ontogenetically patterned in terms of a theory of socialization, led individuals to act in such a way as to generate an isomorphism between cultural and biological success. Which learning preferences? The individual-level learning preferences involved are probably the "reinforcing" and "intrinsically motivating" behaviors studied by the psychologists, but only actual field research could establish which ones: desire for food, or the admiration of others, or the feeling of competence, or whatever.[40] Or perhaps the relevant experiences have reward qualities of which the psychologists in their laboratories are still unaware. New vistas of social psychological field research open. New possibilities of generative theories of sociocultural evolution, based on the learning preferences analyzed by the psychologists, must also emerge. The vertical integration of a human sociobiology demands rather than permits cross-fertilization and cooperation among different disciplines and levels of analysis.

The development of a human sociobiology has already begun, even if in unsystematic fashion. Political science, under ethological influence, is probably the social science best informed by a comparative and at times evolutionary approach.[41] Human ethology, despite its at times excessive concern with young children and the more stereotyped aspects of human behavior, has clearly demonstrated that the ethnographer's conception of an all-powerful culture molding an infinitely malleable human species is very much an exaggeration.[33] A look of coyness is a look of coyness, regardless of the culture: Eibl-Eibesfeldt's photographs establish this more clearly than would reams of computer print-out.[42]

An excellent model for a human sociobiology comes from the study of social hierarchy. Social interaction generates status differential in many species. At the evolutionary level, this phenomenon is readily explained by inclusive fitness theory: high status is at least loosely associated with preferential access to inclusive fitness-enhancing values such as food, or mates. On the other hand, accepting a lower rank after only preliminary threats or a limited, ritualized combat is also a fitness-enhancing strategy, since the alternative would be to risk physical injury or even death in a contest with an opponent more vigorous or with more allies than oneself. It is therefore not surprising that status is a general primate trait: the social orders of young human children bear striking resemblances to those of apes and old world monkeys.[43] But as the human child matures, he increasingly comes to determine relative standing not in terms of agonistic encounters but according to culturally patterned symbolic criteria of social prestige,[44] such as skill in athletic displays or in hunting or farming, or personal wealth or the standing of parents. At the level of personality theory, social hierarchy is intimately related to the maintenance of self-esteem and also to identity, for one's reference groups supply the criteria in terms of which one evaluates one's relative standing. Failure to achieve acceptable standing in terms of the criteria of a particular reference group may lead to low self-esteem, or to a change of reference groups. At the level of culture theory, the prestige criteria of reference groups is often related to the subsistence pattern and ecology of the society. Traditional economic systems generally depend upon individuals striving to be superior hunters, or farmers, or craftsmen, or whatever: they so strive because their identity is bound up with a reference group that will determine their relative standing in terms of how skilled they are at hunting, or farming, and so forth. In effect, individual-level striving for identity, self-esteem, and social prestige motivate the behaviors necessary for the perpetuation of traditional economic systems. Yet the evolutionary basis of status hierarchy is clearly demonstrated by the universality of aspects of its nonverbal communication, e.g., lower physical position and gaze aversion indicate lower status everywhere. The study of social hierarchy is an example of the kind of vertically integrated, multilevel sets of theories a human sociobiology would produce. Note, too, that the various levels of analysis are not separate and independent, even though the organization of academic disciplines gives that impression. In the social hierarchy case, for example, personality-level self-esteem is intertwined with theories of inclusive fitness, social prestige, subsistence economy, ontogeny, and nonverbal communication.

It would have been good to conclude this section with a vertically integrated treatment of human sexuality, since much of the earlier discussion was devoted to that topic. Though I cannot, I do know what such a synthesis would look like. It would begin with the discussion of parental investment already presented but would go on to examine that theory's implications in the light of various likely reconstructions of the course of human evolution. It would discuss sexual behavior across a variety of species, so that we could better understand how the ecology, too, generates selective pressures which interact with those predicted by parental investment theory. With this evolutionary background in mind, it would examine the physiological and neurophysiological mechanisms mediating the

predicted selective pressures. It would look at the range of variation
in sexual behavior, including courtship and bonding, across the
entire species (and not just within a single society), and then assess
the extent to which central tendencies were or were not consistent
with both the evolutionary and physiological theories. Any incon-
sistencies would require a re-examination of the original data and
assumptions that produced the conflicting interpretations. Reviewing
cross-cultural differences in the patterning of sexuality and bonding,
it would relate these data to theories of culture and ecology, and
perhaps provide historical-particularistic interpretations of patterns
not consistent with the general framework. It would predict that
those cultural patterns of sexuality that force individuals to diverge
greatly from the kinds of behaviors predicted by evolutionary theory
should also be those that are distant from the central tendency
(species-typical) cultural patterns: if they are not, then errors at
one or more levels have been made. Assuming all such errors have
been corrected, individuals in such deviantly patterned cultures
should show symptoms of stress, since their cultures are bending
them against the grain of human nature. Finally, a common core
of cross-culturally universal aspects of non-verbally communicative
gestures and voice-tones would be catalogued and their role in
courtship and bonding assessed, along with the emotions underlying them and their possible neurophysiological concomitants. The
result of this constant cross-checking of levels of theory and data
would be a vertically integrated science of sexology, one branch of
a human sociobiology.

Conclusions

Sociobiology in neither its broad nor narrow sense is a theory of
human nature. Broadly defined, it is a discipline studying the behavior of animals and insects, distinguishable by its integrative,
evolutionary perspective. Narrowly defined, it refers to recent and
exciting developments in evolutionary theory whose application to
human behavior is probably premature without careful reconstruction of the ecology and social organization of *Homo erectus* and
early *Homo sapiens*. Yet, sociobiology can give rise to an integrated
study of human behavior and society.

The coming of the biologists to the social and behavioral sciences and their use of a seductively simple but powerful evolutionary perspective have accented the fragmented nature of these fields. Theoretical bits and pieces float about in a sort of pre-paradigmatic sea.[45] Here a piece may be labelled "reference group theory" and there is an elegant bit titled "attribution theory." Here floats a ship of economic determinists and there sail those for whom culture is a system of symbols that orders all. Even the writers of textbooks tend to treat levels of explanation other than their own as irrelevant and "reductionistic," thereby becoming reductionists themselves. An evolutionary framework that recognizes the need for multiple-level yet linked explanations can put it all together. Vertically integrated theory is possible, as we have seen. But if we can have a human sociobiology, will we?

The present disorder of the social-behavioral sciences is as much sociological as historical in nature. Socialization into any one discipline often involves indoctrination against others. Biologists really do tend to feel, privately, that had they a few months free they could reinvent the social sciences and clear away the existing debris. Social/cultural anthropologists and sociologists make simplistic and shaky motivational assumptions and then erect vast theoretical edifices on these weak psychological bases. Psychologists have so strong an anti-evolutionary bias that they have long discussed "reinforcers" without asking why one species should find a stimulus reinforcing and another species fail to respond to it.[46] Students at a major North American university were warned by their sociology professor recently that the lectures to be given by Edward O. Wilson were not worth attending because their "hidden agenda" had to do with biological reductionism at best, racism at worst. Sociobiologists and their critics tend to assume adversary stands, presenting arguments at different levels of analysis as if they were opposed rather than complementary to one another.

The reward system of academic institutions penalizes the scholar who moves to areas of overlap with other disciplines, for his departmental and disciplinary colleagues automatically view his work as of less importance than that which lies in more traditionally central areas. He has fewer allies in the political infighting endemic to academia. He finds it difficult to fit into the standard "slots" of conservative departments. As a result, the disciplines condense around the motes of determined-by-historical-accident "central concerns" and grow increasingly distant from one another. They become islands among which undergraduates find it frustratingly difficult to swim. Those of their professors who respond to the bridge-building goad of intellectual curiosity may thereby weaken their careers (unless they are already very firmly established).[47]

Wilson's *Sociobiology* and the powerful ideas it represents are stirring segments of the academic community. It is forcing a looking outward, beyond disciplinary boundaries. Perhaps it will lead the senior professors who control academic departments and who review grant applications to encourage cross-disciplinary, vertically integrated theory and research. Perhaps it will even inspire a parallel volume, one synthesizing the human behavioral and social sciences into a single discipline united by the framework of evolutionary biology—a human sociobiology.

Summary

Recent developments in evolutionary biology have permitted that field to deal powerfully with the evolution of social behavior. The resulting theoretical approach is often referred to as "sociobiology." When applied to human behavior, this approach gives rise to what at times is almost a theory of human nature. Despite the insights and stimulation it provides, however, sociobiology is not such a theory because: 1) it fails to take into account selective pressures generated by early protohominid environment; and 2) it does not deal with the physiological, individual, or cultural levels of explanation. A theory of human nature and society must explain behavior on at least four levels, and each explanation must be compatible with but not reducible to the other levels. Human behavior at both the individual and cultural levels of organization has emergent properties not reducible to the maximization of inclusive fitness axiom of sociobiology. What is required is a human sociobiology, a vertically integrated discipline in which behavior is dealt with at several levels simultaneously, with special attention to the compatibility of each level with all the others, and to feedbacks among them. Both the reward and training systems of academic disciplines maintain the current fragmentation of the social-behavioral sciences and penalize those who attempt vertically integrated, cross-disciplinary theory and research.

NOTES

1. Richard D. Alexander, "The Search for an Evolutionary Philosophy of Man." *Proc. Royal Soc.* (Victoria) *84*(1971):99–120; Richard D. Alexander, "The Evolution of Social Behavior." *Ann. Rev. Ecol. Syst.* *5*(1974):325–83; Richard D. Alexander, "Natural Selection and the Analysis of Human Sociality." *In,* C. E. Goulden (ed.), *Changing Scenes in the Natural Sciences.* Phil.: Philadelphia Acad. Sci., 1976.
2. David P. Barash, *Sociobiology and Behaviour.* New York: Elsevier, 1977.
3. Richard D. Alexander, "The Search for a General Theory of Behaviour." *Behavioural Science* *20*(1975):77–100.
4. W. D. Hamilton, "The Genetical Evolution of Social Behaviour." *Journal of Theoretical Biology* *7*(1964)1–52; W. D. Hamilton, "Selfish and Spiteful Behaviour in an Evolutionary Model." *Nature* *228*(1970):1218–20; W. D. Hamilton, "Selection of Selfish and Altruistic Behaviour in Some Extreme Models." *In,* John F. Eisenberg and Wilton S. Dillion (eds.), *Man and Beast: Comparative Social Behavior.* Washington, D.C.: Smithsonian Institution Press, 1971, pp. 57–91.
5. W. D. Hamilton, "Innate Social Aptitudes of Man: An Approach from Evolutionary Genetics." *In,* Robin Fox (ed.), *Biosocial Anthropology.* New York: John Wiley, 1975, pp. 133–56.
6. Robert L. Trivers, "The Evolution of Reciprocal Altruism." *Quarterly Review of Biology* *46* (1971): 35–37.
7. Robert L. Trivers, "Parental Investment and Sexual Selection." *In,* Bernard G. Campbell (ed.), *Sexual Selection and the Descent of Man.* Chicago: Aldine, 1972, pp. 136–79.
8. Robert L. Trivers, "Parent-Offspring Conflict." *American Zoologist* *14*(1974):249–264.
9. Robert L. Trivers and H. Hare, "Haplodiploidy and the Evolution of the Social Insects." *Science* *191*(1975):249–63.
10. M. J. West-Eberhard, "The Evolution of Social Behavior by Kin Selection." *Quarterly Review of Biology* *50*(1975):1–34.

11. Edward O. Wilson, *Sociobiology: The New Synthesis*. Cambridge, Mass.: Belknap Press of Harvard University Press, 1975.

12. Richard D. Alexander, "Natural Selection and Social Change." *In*, R. L. Burgess and T. L. Huston (eds.), *Social Exchange in Developing Relationships*. New York: Academic Press, in press.

13. Richard Dawkins, *The Selfish Gene*. New York: Oxford University Press, 1976.

14. G. C. Williams, *Adaptation and Natural Selection*. Princeton, N.J.: Princeton University Press, 1966.

15. P. W. Sherman, "Nepotism and the Evolution of Alarm Calls." *Science,* Vol. 197 (23 Sept. 1977), pp. 1246–53.

16. Indeed, the respected anthropologist M. Sahlins (*The Use and Abuse of Biology: An Anthropological Critique of Sociobiology* [Ann Arbor: University of Michigan Press, 1977]) argues that sociobiology is nothing but the application of capitalist ideology to evolutionary theory and the entire field should not be taken seriously by social scientists.

17. Jeffrey A. Kurland, "Paternity, Mother's Brother, and Human Sociality." *In*, N. A. Chagnon and W. Irons (eds.), *Evolutionary Biology and Human Social Behavior: An Anthropological Perspective*. North Scituate, Mass.: Duxbury, 1979, pp. 145–80.

18. See, for example, Robert A. LeVine and Donald T. Campbell, *Ethnocentrism and Theories of Conflict*. New York: John Wiley, 1972.

19. Norbert Bischof, "Comparative Ethology of Incest Avoidance." *In*, Robin Fox (ed.), *Biosocial Anthropology*. New York: John Wiley, 1975, pp. 37–68.

20. For fuller discussion of the implications of Trivers's ideas for human sexuality and sex differences, see Roger R. Larsen, "Les Fondements évolutionnistes des différences entre les Sexes." *In*, Evelyne Sullerot (ed.), *Le Fait féminin*. Paris: Payard, 1978, pp. 337–358; and Jerome H. Barkow, "Evolution et Sexualité." *In*, Claude Crépault and Joseph Lévy (eds.), *La Sexualité humaine: Textes fondamentaux*. Montréal: Presse de l'Université du Québec, n.d.

21. Napoleon A. Chagnon, "Is Reproductive Success Equal in Egalitarian Societies?" *In*, N. A. Chagnon and William Irons (eds.), *Evolutionary Biology and Human Social Behavior: An Anthropoligical Perspective,* North Scituate, Mass.: Duxbury, 1979, pp. 374–401.

22. Jerome H. Barkow, "Conformity to Ethos and Reproductive Success

in Two Hausa Communities: An Empirical Evaluation." *Ethos* 5 (1977) 409–25.

23. Jerome H. Barkow, "Biological Evolution of Culturally Patterned Behavior." *In,* Joan Lockard (ed.), *Evolution of Human Social Behavior,* New York: Elsevier, in press.

24. This discussion is summarized from Jerome H. Barkow, "Culture and Sociobiology." *American Anthropologist* 80(1978):5–20.

25. Jerome H. Barkow, "Causal Interpretation of Correlation in Cross-Cultural Studies." *American Anthropologist* 69 (1967):506–10; Charles Harrington and John W. M. Whiting, "Socialization Process and Personality." *In,* Francis L. K. Hsu (ed.) *Psychological Anthropology,* new edition. Cambridge, Mass.: Schenkman, 1972, pp. 469–508.

26. Jerome H. Barkow, "Social Norms, the Self, and Sociobiology: Building on the Ideas of A. I. Hallowell." *Current Anthropology* 19(1978):99–118.

27. Nicholas G. Blurton Jones, "Growing Points in Human Ethology: Another Link Between Ethology and the Social Sciences?" *In,* P. P. G. Bateson and R. A. Hinde (eds.), *Growing Points in Ethology,* Cambridge: Cambridge Univ. Press, 1976, pp. 427–50.

28. John A. Bowlby, *Attachment and Loss. Vol. I: Attachment,* London: Hogarth, 1969; Robert A. Hinde, *Biological Bases of Human Social Behavior,* New York: McGraw-Hill, 1974; Lynn A. Fairbanks, "Animal and Human Behavior: Guidelines for Generalization Across Species." *In,* Michael T. McGuire and Lynn A. Fairbanks (eds.), *Ethological Psychiatry: Psychopathology in the Context of Evolutionary Biology.* New York: Grune & Stratton, 1977, pp. 87–110.

29. Jerome H. Barkow, "Attention Structure and Internal Representations." *In,* Michael R. Chance and Raymond R. Larsen (eds.), *The Social Structure of Attention.* London: Wiley, 1976, pp. 203–19.

30. Representative discussions of the selective pressures affecting early human behavior are to be found in (29) and G. E. King, "Socio-territorial Units Among Carnivores and Early Hominids," *Journal of Anthropological Research* 31(1975):69–87; M. F. A. Montagu (ed.), *Culture and the Evolution of Man.* New York: Oxford University Press, 1962; A. Roe and G. G. Simpson (eds.), *Behaviour and Evolution: A Symposium.* Chicago: Univ. of Chicago Press, 1960; S. L. Washburn (ed.), *Social Life of Early Man.* New York: Viking Fund Publications, 1961; and J. N. Spuhler (ed.), *The Evolution of Man's Capacity for Culture.* Detroit: Wayne State Univ. Press, 1959.

31. L. D. Mech, *The Wolf*. Garden City, N.Y.: Natural History Press for the American Museum of Natural History, 1970.

32. L. von Bertalanffy, *General System Theory*. New York: George Braziller, 1968.

33. Jerome H. Barkow, "Human Ethology and Intra-Individual Systems," *Social Science Information 16*(1977):133–45.

34. For example, see J. W. Bennett, *The Ecological Transition: Cultural Anthropology and Human Adaptation*. New York: Pergamon Press, 1977.

35. Robert B. Edgerton, *The Individual in Cultural Adaptation: A Study of Four East African Peoples*. Berkeley: Univ. of California Press, 1971.

36. As Dr. I. Charles Kaufman points out, in a personal communication, rigidly specifying levels of analysis risks falling back into the hole from which the ladder of vertical integration permits us to emerge. Indeed, only in a metaphorical sense is a physiological explanation at a different "level" than an analysis of behavior in terms of personality dynamics. The point is that behavioral phenomena are multi-determined and thus require multiple but mutually compatible explanations.

37. Jerome H. Barkow, "Darwinian Psychological Anthropology: A Biosocial Approach." *Current Anthropology 14*(1973):373–88.

38. See the "Reply" in (26), p. 115.

39. William Irons, "Culture and Biological Success." *In*, N. A. Chagnon and W. Irons (eds.), *Evolutionary Biology and Human Social Behavior: An Anthropological Perspective*. North Scituate, Mass.: Duxbury, 1979, pp. 257–72.

40. For example, see E. L. Deci, *Intrinsic Motivation*. New York: Plenum, 1975.

41. For examples, see A. Somit (ed.), *Biology and Politics*. The Hague: Mouton, 1976; or Fred H. Wilhoite, Jr., "Primates and Political Authority: A Biobehavioral Perspective." *American Political Science Review 70*(1976):1110–26.

42. I. Eibl-Eibesfeldt, *Ethology*. 2nd ed. New York: Holt, Rinehart & Winston, 1975.

43. See the various chapters in D. R. Omark, D. G. Freedman, and F. Strayer (eds.), *Dominance Relations: Ethological Perspectives on Human Conflict*. New York: Garland, *forthcoming;* see also the research review by R. C. Savin-Williams and D. G. Freedman, "Bio-Social Approach to Human Development." *In*, Suzanne Chevalier-Skolnikoff and Frank E. Poirier (eds.), *Primate Bio-Social Development: Biological, Social, and Ecological Developments*. New York: Garland, 1977, pp. 563–602.

44. Jerome H. Barkow, "Prestige and Culture: A Biosocial Interpretation." *Current Anthropology* 16(1975):553–72; Jerome H. Barkow, "Strategies of Self-Esteem and Prestige in Maradi (Niger Republic)." *In,* Thomas R. Williams (ed.), *World Congress: Psychological Anthropology.* The Hague: Mouton, 1975, pp. 373–88.
45. The description of the social sciences as "pre-paradigmatic" is taken from T. S. Kuhn, *The Structure of Scientific Revolutions.* Chicago: Univ. of Chicago Press, 1962.
46. Robert B. Lockhard, "Reflections on the Fall of Comparative Psychology: Is There a Message for Us All?" *American Psychologist* 26(1971):168–179.
47. This discussion owes much to two papers, that of D. T. Campbell, "Ethnocentrism of Disciplines and the Fish-Scale Model of Omniscience." *In,* Muzafer Sherif and Carolyn W. Sherif (eds.), *Interdisciplinary Relationships in the Social Sciences.* Chicago: Aldine, 1969, pp. 328–48; and that of Murray Wax, "Myth and Interrelationship in Social Science: Illustrated Through Anthropology and Sociology," in the same volume, pp. 77–99.

SOCIOBIOLOGY: THE "ANTIDISCIPLINE" OF ANTHROPOLOGY

When *The Insect Societies* was published in 1971 it contained a final chapter on "the prospect of a unified sociobiology" in which E. O. Wilson envisaged that "the same principles of population biology and comparative zoology" which had "worked so well in explaining the rigid systems of the social insects could be applied point by point to vertebrate animals." And he then went on to aver (in homiletic words taken from a French text of 1810 on ants) that "the most dignified end of which science may boast" is "endeavoring to ameliorate the human species with the examples its lays before us." That same year, E. O. Wilson (1971b:209) had dwelt on our "grave need for a truly scientific and powerful anthropology." What, one wondered, with trepidation, might be the rigidifying "examples" that sociobiology had in store for anthropology's amelioration.

The magniloquent answer came in 1975 in the perfunctory final chapter of Wilson's monumental tome in which sociobiology was unveiled as "The New Synthesis." Looked at macroscopically, and "in the free spirit of natural science," Wilson proclaimed (p. 547), "the humanities and social sciences shrink to specialized branches of biology." Indeed, "one of the functions of sociobiology" (p. 3) was "to reformulate the foundations of the social sciences" so that they became "truly biologicized." When thus ameliorated anthropology (together with sociology) would constitute (p. 574) "the sociobiology of a single species."

This heady vision of anthropology's engulfment by sociobiology stems directly from Wilson's reductionist suppositions. Biology, Wilson believes, is the "antidiscipline" of the social sciences, and, of anthropology, in particular, which (1977:134) "has already become the social science closest to sociobiology." It is, Wilson argues, the scientific function of an "antidiscipline" to reduce to its own level the explanations of the science next above it; and so, it falls to sociobiology, as the "antidiscipline" of anthropology, to demonstrate that the explanation of cultural practices is really to be found in biological processes. Or, to state the situation more specifically, to show that the explanations of the actions of humans in any culture can be reduced to the calculus of inclusive fitness (Barash, 1977:327), it being the basic principle of sociobiology that an organism's evolutionary success is measured by the extent to which its genes are represented in the next generation.

In his anthropological essay of 1978, *On Human Nature,* in which he elaborates the conclusions summarily stated in the final chapter of *Sociobiology,* Wilson strongly maintains this reductionist stance. Sociobiology remains the "antidiscipline" of anthropology, and (p. 2), "the essence of the argument" is "that the brain exists because it promotes the survival and multiplication of the genes that direct its assembly."

In what follows I shall argue that sociobiological theory, however great be its explanatory power when applied to species the closed behavior programs of which are genetically determined, will not suffice in the case of *Homo sapiens,* and will, accordingly, never amount to an adequate anthropology, or human ethology.

I would emphasize that in so arguing, I accept the same "scientific materialism" as that in which E. O. Wilson places his trust, as also the theory of evolution by means of natural selection in its modern form. Indeed, I hold it to be of the utmost importance that anthropologists should recognize and incorporate in their theories all those biological mechanisms which have been shown, on scientific grounds, to be relevant to the understanding of human behavior. It is very much my view, then, (Freeman, 1974:221) that "the modern theory of evolution" is basic to "an authentic science of anthropology." But equally, I do not suppose that acceptance of the theory of evolution by means of natural selection is in any way incompatible with the full recognition of the learned behavior and symbolic systems by which human populations have long been characterized, and which depend on the transmission of information by other than genetic means.

The issue at stake is the extent to which human cultures, and the behaviors which are a part of them, can be accounted for by "genetic determinism." It is on the interpretation of this "key phrase," as Wilson (1978:55) stresses, that "the entire relation between biology and the social sciences" depends.

As an example of an organism whose behavior is genetically "predestined," Wilson instances the mosquito. "The mosquito," he writes (1978:55), "is an automaton," with "a sequence of rigid behaviors programmed by the genes to unfold swiftly and unerringly from birth to the final act of oviposition."

He then immediately goes on to take as an example of "restricted behavior" in humans the phenomenon of handedness. "Each person," he writes (p. 57), "is biologically predisposed to be either left- or right-handed." Yet, as Wilson notes, while parents in present-day Western societies tend not to interfere with "the direction set by the gene affecting this trait," the position is decidedly different in "traditional Chinese societies," in which "a strong social pressure for right-handed writing and eating" is still exerted. Thus, in a recent study of Taiwanese children, Evelyn Lee Teng and her associates (1976:1146) "found a nearly complete conformity in these two activities but little or no effect on handedness in other activities not subjected to special training" (Wilson, 1978:57).

The conclusion that Wilson draws from this instructive example is that in the case of the "restricted behavior" of handedness "the genes have their way unless specifically contravened by conscious choice." What it also plainly demonstrates is that among humans the actions of individuals may be determined not by the relevant pre-existing genetic program, but instead, by an alternative which, in the course of the history of the population concerned, has come to be culturally preferred and socially sanctioned. In such a case, I shall argue, the alternative action is a direct expression of a highly developed capacity to make choices which, when it is viewed in evolutionary perspective, is seen to be one of the defining characteristics of the human ethogram.[1] Moreover, as I shall also argue, it is precisely because of this marked human capacity for non-genetically determined alternative action that sociobiological theory, when applied to human populations, is irremediably deficient.

In the two instances cited by Wilson, that of the mosquito and that of Taiwanese children, we are dealing with what Mayr has called closed and open programs of behavior. This crucial distinction Mayr has described as follows (1976a:23):

> The young in some species appear to be born with a genetic program containing an almost complete set of ready-made, predictable responses to the stimuli of the environment. We say of such an organism that its behavior *program* is *closed*. The other extreme is provided by organisms that have a great capacity to benefit from experience, to learn how to react to the environment, to continue adding "information" to their behavior program, which consequently is an *open program*.

The term "program," as Mayr notes, (1976b(*orig.* 1974):393) is derived from the language of information theory, and, as he uses it, refers to "coded or prearranged information that controls a process (or behavior) leading it towards a given end" (cf. also, Young, 1978:8). More specifically then, a behavior program may be said to be *closed* (Mayr, 1976c:696) when "nothing can be inserted in it through experience," and *open* when it "allows for additional input during the lifespan of its owner."

In nature we find species with closed behavior programs, like the mosquito, thriving alongside species with open programs, such as *Macaca fuscata*, the Japanese macaque, or *Homo sapiens*. It is thus evident that both kinds of programs can be highly adaptive. Mayr also presents an explanation of why each of these widely differing forms of adaptation should have evolved.

> Since much of the behavior directed toward other conspecific individuals consists of formal signals and of appropriate responses to signals [he writes (1976c:708)], and since there is a high selective premium for these signals to be unmistakable, the essential components of the phenotype of such signals must show low variability and must be largely controlled genetically.

In other words, selection tends to favor the evolution of a closed program "when there is a reliable relationship between a stimulus and only one correct response."

In contrast, behavior leading to an exploitation of natural resources has to be flexible to make possible "an opportunistic adjustment to rapid changes in the environment," the enlargement of a niche, or the shift to a new one. Such flexibility could not develop, however, if this behavior were (p. 708) "too rigidly determined genetically." What happened then, during the course of evolution in the case of some species, was "a gradual opening up of the genetic program, permitting the incorporation of personally acquired information to an ever-greater extent."

In Mayr's view there are certain prerequisites if this gradual opening up of a genetic program is to occur. Because personally acquired information necessitates "a far greater storage capacity than is needed for the carefully selected information of a closed genetic program" a larger central nervous system is required. Again, the development of an open program is favored by prolonged parental care, for "when the young of a species grow up under the guidance of their parents they have a long period of opportunity to learn from them," and are thus able to fill their open programs with "useful information on enemies, food, shelter, and other important components of their immediate environment."

In summary then, it can be said that an open program of behavior is one that does not prescribe all of the steps in a behavioral sequence, but (Popper, 1977b) "leaves open certain alternatives, certain choices, even though it may perhaps determine the probability or propensity of choosing one way or another."

As Mayr, Popper, and Young all emphasize, both open and closed programs of behavior have been evolved by means of natural selection. There is now, however, conclusive evidence to show that an open program, itself the product of natural selection, can, in some species, lead to the formation of what are called *traditions;* and when this happens (as has been observed in several primate species) we are dealing with the initial stage of a mode of transmitting information from generation to generation within populations that is no longer directly dependent (as are closed programs of behavior) on the operation of a genetic code.

That this particular kind of evolutionary development has now been scientifically verified is of the utmost significance for the understanding of the special character of human history, which, at least from the time of *Homo habilis* (about 1.75 million years ago) has involved both the genetic and the exogenetic inheritance of information.

Wilson (1975:168) defines traditions as "specific forms of behavior that are passed from generation to generation by learning," and notes their presence in the behavioral repertoires of a wide range of animal species. In particular, he refers to the "carefully documented case histories" of the formation of traditions that have come from studies of the Japanese macaque (*Macaca fuscata*), describing how, from 1952 onwards when they were experimentally provisioned, new modes of food gathering and handling, originating in the innovative actions of an individual member, became established by way of observational learning, throughout most of the Koshima Island troop, and were then transmitted, by this same means, from mothers to infants.

In Wilson's view, "the innovations of the Koshima troop" (p. 171) provide "a graphic illustration" of the potential role of learned behavior or tradition as "an evolutionary pacemaker."

> The food presented to the monkeys on the beach [Wilson comments] attracted them to a new habitat and presented them with opportunities for further change never envisioned by the Japanese biologists. Young monkeys began to enter the water to bathe and splash, especially during hot weather. The juveniles learned to swim, and a few even began to dive and to bring up seaweed from the bottom. One left Koshima and swam to a neighboring island. By a small extension in dietary opportunity the Koshima troop had adopted a new way of life, or more accurately, grafted an additional way onto the ancestral mode.

"It is not too much," Wilson adds, "to characterize such populations as poised on the edge of evolutionary breakthroughs. . . ."

In these "carefully documented case histories" of the Japanese macaque we have a striking example of the way in which the members of a primate species, bequeathed by natural selection with an open program of behavior, have the capacity, when presented with the opportunity, to take up a behavioral alternative that was not a part of their immediately pre-existing ethogram.

However, Wilson's depiction of this process as one in which an "additional way" is "grafted" on to an "ancestral mode," obviously needs further analysis. As Wilson himself stipulates, the "additional way" we are discussing is, in fact, a newly acquired tradition, or socially learned alternative. Further, the process by which this newly acquired tradition is "grafted" on to the ancestral ethogram of the species concerned is not directly through the mechanisms of its genetic code, but instead, by the effecting of changes in the neuronal structure of the brains of the individuals involved so that an adequate memory of a behavioral alternative is laid down.

In other words, we are dealing with a process in which information is being transmitted between animals by observational learning and stored in their brains by specifically nongenetic mechanisms.

Thus, while the brains of learning animals synthesize RNA and protein, and while, as Blakemore expresses it (1977:114) "the protein produced is derived ultimately from messages in the genetic DNA, it is not *that* code that contains the new, personal memory." Instead, each memory is coded, in a way not yet fully understood, by mechanisms that operate within the neurones of the brain.

Thus, while it is certain that these mechanisms which are the essential prerequisites for an open program of behavior have been evolved by natural selection, it is equally true that they are also mechanisms which, in their operation, do *not* directly involve the genetic code. To provide the genetic basis for an open program of behavior and the formation of traditions, all that the chromosomes have to store, to quote Blakemore once again (p. 114), are "the instructions needed to allow any circuit to change depending on its own activity, without specifying in advance which circuit will be involved."

What the selective pressures were that led to the emergence of the capacity to learn, and so form traditions, I shall not here conjecture. Whatever these pressures may have been they most certainly set in train, in various animal species, exogenetic processes of information transmission that have had momentous consequences, and most markedly of all in the unparalleled history of our own species. There are good grounds then, for Blakemore's referring to the emergence of the capacity to learn as a "triumph of evolution." "Its first appearance," as he remarks, "must have been, quite simply, a transcendent step in the development of animal life, for learning frees the individual from the chains of his own double helix."

There is then, abundant evidence, as Wilson himself documents (1975:168), that populations of primates and other animals possess traditions, or "specific forms of behavior that are passed from generation to generation by learning," by processes, that is, which are, specifically, exogenetic. These traditions, when viewed collectively, constitute the *culture* of an animal population, and, as Wilson notes (1975:539), from the researches of primatologists during recent decades, it has now been established that "the rudiments of culture are possessed by higher primates other than man, including the Japanese monkey and the chimpanzee."

For populations of both the Japanese macaque and the chimpanzee, then, we now possess decisive evidence of the exogenetic transmission of information between successive generations.[2] The fact that the information transmitted is rudimentary in form, is, moreover, when seen in evolutionary perspective, of particular significance, for in both of these meticulously documented cases we are given a glimpse of the way in which the cultural transmission of information emerges, through the operation of entirely natural processes, from pre-existing genetically determined adaptations.[3]

The evidence we possess also indicates that the evolutionary emergence of the cultural transmission of information in all species, including Man, is an exceedingly gradual process, with an unbroken temporal continuity in the interaction of the pre-existing genetic and emerging exogenetic code, even though in the subsequent course of their evolutionary histories these two codes become quite distinct in their separate functions.

Given the demonstrated presence in some infra-human primate populations of the incipient exogenetic transmission of information, in addition to its long-standing genetic transmission, we clearly have the empirical evidence to make a general distinction between these two codes, and to argue that their co-existence, during several million years of history, has, in the singular case of the human species, resulted in the emergence of two separate, though very closely interacting, evolutionary systems.[4]

The principal significance for anthropology, then, of the knowledge we now possess about the rudimentary cultures of chimpanzee and Japanese macaque populations is that it enables us to infer that culture-forming behavior of a comparable kind was characteristic of the early hominids of, say, 4 million years B.P., and, probably, even for millions of years before that time, as a rudimentary expression of their open behavior programs.

It is, however, only with the appearance of stone tools, 1.75 million years and more ago, that we can be certain of the presence in hominid populations of the capacity to create and maintain traditions. As Clark and Piggot (1970:33) have observed, "Man's earliest essays in culture are best traced through his artifacts of flint and other kinds of stone," which they appropriately term "cultural fossils." This concept of the "cultural fossil" is highly pertinent to my theme, for it refers not to a fossil that can be classed as the product of the 1st, or organic evolutionary system, but, instead, to an artifact produced by the innovative behavior of a hominid, or, to be more precise, to their instrumental choice behavior, for such artifacts can only have resulted from the selection, by their makers, of certain specific alternatives. These "cultural fossils" then, are indisputable evidence of the existence, in the pre-*sapiens* populations of the Lower Paleolithic, of a 2nd evolutionary system characterized by the intergenerational transmission of exogenetic information.

Thus, while the genetic evolution of these populations was actively continuing, it was accompanied, from this time onwards, by a 2nd evolutionary system based on different mechanisms; and so, from at least the beginnings of their Paleolithic periods, all populations of the genus *Homo* have, to use Wilson's own words (1978: 78) moved on "a dual track of evolution," with two clearly distinguishable evolutionary systems—the first genetic and the second cultural—in constant interaction.

Furthermore, during the last 40,000 years or so of this still continuing "dual track of evolution," there has been a marked cumulative change in the relationship of its dual systems, with cultural evolution assuming an ever increasing importance in human history.

Washburn has recently suggested (1978:154) that, while man "was surely not mute for most of his development," the "critical new factor," that provided "a biological base" for "the acceleration of history" from about 40,000 years ago onwards, was "the development of speech as we know it today." This view is very much in accord with the conclusions of Isaac (1976:275), who infers, on archaeological grounds, that while "the milieu in which capabilities for language were first important" began more than a million years ago, the crucial developments in language took place about 30,000 to 40,000 years ago; and, it would seem likely, as Washburn has proposed, that "just as upright walking and toolmaking were the unique adaptation of the earlier phases of human evolution, so was the physiological capacity for speech the biological basis for the later stages."

Of the crucial importance of effective languages in these later stages of human prehistory there can be no doubt. What a spoken language provides, with its uniquely human phonetic code, is an extraordinarily potent means for generating new information, a feature which has been well described by Wilson (1975:555):

> In any language words are given arbitrary definitions within each culture and ordered according to a grammar that imparts new meaning above and beyond the definitions. The full symbolic quality of the words and the sophistication of the grammar permit the creation of messages that are potentially infinite in number.

This means that all human languages possess what Steiner (1976:228) has called alternity; that is, they immensely facilitate the conceptualizing of possibilities not previously perceived and so generate new alternatives from which choices can be made.

With the development of effective languages then, the 2nd evolutionary system of human populations was transformed, for it was now possible to supplement the observational learning that first gave rise to traditions, with a highly efficient symbolic code in which cultural information of all kinds could be stored and transmitted from generation to generation. Man had become a *zoon phonanta,* or language animal, and from the time of the effective completion of this transformation, about 40,000 years ago, his evolutionary history has been mainly cultural.

The anthropological significance of this final stage of Man's gradual transition from a preponderantly genetic to a preponderantly cultural mode of evolution is difficult to exaggerate, and Wilson may well be justified in claiming (1975:556) that the development of human speech represented "a quantum jump in evolution comparable to the assembly of the eucaryotic cell."

From about 40,000 years B.P. onwards, certainly, the advanced Paleolithic peoples, equipped, we must suppose, with effective languages, began to explore an ever-extending range of alternatives, and, toward the end of the Paleolithic, there was a sustained flowering of human inventiveness and creativity evinced in devices like the spear-thrower, the harpoon, and the eyed-needle, as well as in multifarious works of art, which ranks, as Jacquetta Hawkes (1976:21) has remarked, as "one of the most astonishing chapters" in all human history. And this was followed, from the Neolithic period onwards, by a continuing efflorescence of human agency in which an astonishingly wide range of alternatives (Hawkes, 1976:61 seq.) were taken up as, throughout the Near East as elsewhere in the world, hundreds of differing cultures assumed new forms.

By the end of the 4th millennium B.C., with the appearance of urban civilizations and the invention of writing, the human capacity for alternative action was producing an unending stream of innovative traditions, and, with understandable hyperbole, Clark and Piggot (1970:328) have described this extraordinary epoch of cultural change as "infinitely variable."

Again, in our earliest historical sources, there is copious evidence of immense variability in the customary and moral behavior to be found in the differing cultures of the 1st and 2nd millennia B.C. Herodotus, for example, records (1858:187) how Darius (521-486 B.C.), having ascertained from a group of Greeks (who practised cremation) that they would not for any sum "feed upon the dead bodies of their parents," then introduced them to some Callatians, whose sacred custom this was, and who, when asked if they would consent to the burning of their dead were dreadfully perturbed.

Custom, in Herodotus' view, as in Pindar's, is "the king of all men." We now know that traditions, or customary practices, are to be found in profusion in all human societies, and, further, as with human action in general, that these alternative behaviors, despite the fact that there is no marked variation in human neuro-physiology from population to population, are exceedingly diverse in character. Moreover, anthropological research has also shown that ideas and customary practices can change with extreme rapidity, as when a group is converted from one religion or belief system to another, or follows the directives of a political or other leader.

Because of the scale of these phenomena and the limited time span in which they occur, the boundless diversity and often extreme variableness of action as revealed in human history, since, say, 5000 B.C., cannot possibly be explained by changes in gene frequencies.

It is thus evident that in the case of the human species, we are dealing with processes that lie beyond the reach of any purely genetic theory of evolution, and that, from at least the beginning of the Paleolithic, human evolutionary history can only be adequately comprehended by a specifically anthropological theory which recognizes two interacting systems, the second of them having emerged (from the pre-existing system of genetic evolution) to assume, from about 40,000 B.P. onwards, an ever-increasing importance to become the principal source of change in all human societies.

Although it is still far from being fully understood, it has long been apparent to most anthropologists, as well as to many biologists,[5] that man's 2nd, or cultural, system of evolution is characterized by processes fundamentally different from those of genetic evolution.

Thus, the information transmitted in cultural evolution is, specifically, exogenetic (cf. Medawar, 1976:502), having been generated, at some time in the past, by human agency.

Again, in the great majority of instances, this information is carried in a linguistic code, and its transfer from generation to generation, which involves teaching, learning, and memory, is also, specifically, exogenetic.

In preliterate populations, cultural information is stored in the memories of their members. In literate societies, however, there are numerous devices, such as books, that permit external storage. The sum total of all this information, which constitutes the cultural heritage of a people, has, by Popper (1976:181), been termed World 3, to distinguish it from the world of physical objects, both inorganic and organic, which he terms World 1, while human agency he calls World 2.

Because it depends on the transmission and storage of exogeneti-

cally coded information by learning, memory, and related exogenetic processes, cultural evolution is basically Lamarckian in form.
In other words, its mechanisms facilitate the social inheritance, from
generation to generation, of information acquired in the course of
experience.

This means, as Dobzhansky (1963:312) has put it, that "cultural modifications can . . . be passed to, potentially, any number
of individuals, irrespective not only of their descent relationships,
but also of space and time."[6] It is this kind of social inheritance of
information (which the mechanisms of genetic evolution do *not*
allow) that accounts for the rapid and cumulative nature of cultural evolution.

When compared with the "still on-going genetic evolution" on
which it is "superimposed," cultural evolution is thus seen to be of
"a fundamentally new type" (Tinbergen, 1978:232). "Just as individual memory has partly released each animal from the immediate restrictions of the genetic code," writes Blakemore (1977:116),
"so the sharing of learned ideas by social animals has added an entirely new dimension to the progress of evolution." And, being
especially well developed in man, this process of shared learning,
as Blakemore points out, has resulted in a "social inheritance of
culturally acquired characteristics," which Flew (1975:102) correctly judges to be "the most distinctive and powerful peculiarity of
our species, as compared with all others."

But it is equally true that in addition to being accumulated in a
Lamarckian way, exogenetic information, in the course of cultural
evolution, is also steadily subjected, by human judgment, to active
selection. In directing attention to this process during a lecture to
the Collège de France in November 1967, Jacques Monod (1969:
16) observed: "A transmittable idea constitutes an autonomous
entity . . . capable of preserving itself, of growing, of gaining in
complexity; and it is therefore the object of a selective process, of
which modern culture is the current but in every way evolving
product."

Popper (1974:108) has also emphasized this characteristic of
cultural evolution, observing that it takes place "outside ourselves,
exosomatically,[7] by the means of the growth of our theories in
World 3"; this being because men (Popper, 1977a:459) are capa-

ble of "conscious critical rejection." This means that, in an incep-
tive way, "cultural evolution is also Darwinian," with "conscious
critical rejection" having a function similar to that which natural
selection performs in genetic evolution.

This conclusion is of especial interest in the light of recent studies
of culture formation in Man and other primates (such as the Japanese macaque), which have revealed that cultures are essentially
interrelated accumulations of chosen alternatives. In the diverse
and variable cultures of *Homo sapiens,* it becomes evident that we
are dealing with alternatives which have been selected by human
agency, and, this being so, any of these alternatives is always open
to modification by the further exercise of this same mechanism.

Human cultures, in other words, are fundamentally of Man's
own devising, and are made possible by the virtually unlimited
capacity of the human brain for alternative action, this capacity being the product of the interaction, for at least several million years,
of biology and culture.

In anthropological perspective then, Man is seen as a primate
whose brain, because his remote ancestors possessed open behavior
programs and developed rudimentary cultures, has undergone an
evolution decisively different from that of any other animal species. The human animal is thus a very special kind of primate.
We are, as Washburn (1963:528) has put it, "the product of
the new selection pressures that came with culture"; and it was
these selection pressures that eventually produced the brain that has
been able, with the development of effective languages, continuously to elaborate cultures as complex systems of symbolically
coded and exogenetically transmitted information.

The human brain is thus adapted, by its unique evolution, to
enact not only the biological programs of our primate physiology,
but also to enact and modify cultural and other humanly devised
alternatives. It is in these terms, therefore, that any anthropologically
adequate definition of human nature must be couched.

Sociobiology is defined by Wilson (1975:595) as "the systematic study of the biological basis of all social behavior," and such
study is obviously scientifically valid in the case of mosquitoes or
any other species the behavior programs of which are genetically
prescribed. But, when it comes to the human species, with its dual

track of evolution and its partly genetic and partly exogenetic behavior programs, sociobiological theory, I would argue, is faced with an insuperable problem. It is this: How can a theory that proceeds entirely in terms of genetic processes possibly take adequate account of the cultural variables in human social behavior which anthropological research has demonstrated to be exogenetic in character?

The only intelligent response open to a sociobiologist in this situation is, I would suggest, to take cognizance of these exogenetic variables, as does Richard Dawkins in his book *The Selfish Gene,* and to recognize (1976:205) that "For an understanding of the evolution of modern man we must begin by throwing out the gene as the sole basis of our ideas on evolution."

Why then do other sociobiologists persist in their inevitably vain attempts to reduce the specifically exogenetic information and processes of culture to the specifically genetic information and processes of biology, when what is plainly required is informed study of the way in which the interaction of cultural and genetic processes has produced the human brain and language and made possible the unique character of human history?

The answer, I would suggest, is adumbrated in another of Dawkins's disarming statements. "We biologists," he has written (1976: 208), "have assimilated the idea of genetic evolution so deeply that we tend to forget that it is only one of the many possible kinds of evolution."

This epistemological bias stems from an exaggerated preoccupation with "the blind decision-making process of natural selection" (Wilson, 1978:197), and from the erroneous supposition that the operation of this process alone, adequately accounts for the whole of human evolution, including the evolution of human cultures. For sociobiologists and others who have come to think in this way, open behavior programs, learning, and exogenetic codes pose no special problem, for, as Slobodkin (1978:234) has noted, "there is a persistent feeling that the range of observed behavior is kept within bounds by natural selection and that any observed behavior is almost certainly adaptive."[8]

Indeed, even when it comes to human cultures, which are exogenetic structures of immense complexity, so blind is the faith of some sociobiologists in the pervasive potency of natural selection, that the mere evocation of this process becomes a substitute for rational argument and the detailed presentation of scientific evidence, the assumption being that a process as powerful as natural selection cannot but be operating.

Barash, for example, having declared at the outset of his book *Sociobiology and Behavior,* which has a commendatory Foreword by E. O. Wilson, that (1977:2) "evolutionary theory is a predictive and analytic tool of enormous power," does not include in his discussion (of 48 pages) of "the sociobiology of human behavior" any kind of specification of the mechanisms by which cultural practices, in general, could conceivably be genetically determined. Instead, there are further paeans (p. 318) to "the ultimate generative force of evolution" acting through natural selection "to maximize the inclusive fitness of individuals" and, in lieu of precise and rational explanations, inchoate talk of (p. 282) "an ultimate biological reason" for the existence of "cultural factors" and of how (p. 324) "our deep evolutionary 'sweet receptors' predominate over our rationality."

Was there ever such pseudoscientific flummery? The theory of sociobiology is predicated on the efficacy of "the blind decision-making process of natural selection" in genetic evolution, and this efficacy I accept. When, however, without due knowledge or reflection, it is crudely and reductively applied to the explanation of human cultures, sociobiology becomes slipshod, null, and ludicrous, a science "with its own internal lightning blind."[9]

Indeed, when without recognition of the exogenetic character of cultural processes, natural selection, operating "to maximize the inclusive fitness of individuals," is evoked as a general explanation of "cultural factors," sociobiology becomes an actively unscientific doctrine, perilously close to that "enthusiastic misapplication of not fully understood genetic principles in situations to which they do not apply" that the Medawars (1977:38) have called *geneticism.*

It becomes important then to realize that some sociobiologists view human cultures through a lens of their own making, and that, for example, when Wilson (1975:539) writes that "only in man has culture thoroughly infiltrated every aspect of life" or that

(1976:343) "human social evolution is more cultural than genetic," these are not the interpretations of a cultural anthropologist
but of a genetic determinist who is deeply convinced (1978:167)
that "the genes hold culture on a leash," and who believes (1978:
207) that "there is reason to entertain the view that the culture of
each society travels along one or the other of a set of evolutionary
trajectories whose full array is constrained by the genetic rules of
human nature."

These are far-reaching substantive claims, yet, as in the case of
Barash, Wilson nowhere gives any kind of coherently detailed account of the way in which genetic processes could conceivably produce, in all their detail and diversity, the endless range of exogenetic phenomena resulting from human action, which are to be
found in the hundreds of differing cultures known to anthropology.
Instead (Wilson 1975:559), we are offered the equivocatingly
vague assertion that while "ethnographic detail is genetically under-
prescribed," this "underprescription does not mean that culture has
been freed from the genes." Any demonstration of the "genetic
prescription" of any actual item of "ethnographic detail" is conspicuously lacking.

The notion of the genetic prescription of behavior to which Wilson appeals is precise and understandable. It refers to the phenomenon of behavior being directly determined by genetically coded
information, as in the case of the mosquito (Wilson 1978:55)
with its closed sequences of "rigid behaviors programmed by the
genes." There is, however, no comparably infrangible genetic prescription of the observable range of human behavior. As we have
seen from Wilson's own example, even in instances where there is
genetic prescription to some degree, as in "handedness" (Wilson,
1978:57), the genes do not "have their way" when "specifically
contravened by conscious choice."

The "conscious choice" to which Wilson is here referring, it is
important to note, is in no sense genetically determined, being instead a culturally chosen alternative that runs directly counter to
the genetically prescribed behavior to which it is applied. What this
example demonstrates indeed is the remarkable potency and also the
error-fraught potentialities of the human capacity for nongeneti-
cally prescribed alternative action.

Similarly, possessing this highly efficacious capacity to make choices, humans may impose conventional meanings on genetically prescribed behaviors. For example, as well as establishing the eyebrow flash as a genetically prescribed or fixed behavior pattern in Man, Eibl-Eibesfeldt's researches (1975:472) have revealed that while the Samoans of Western Polynesia have selected the alternative of giving the eyebrow flash the connotation of "approval" as a "factual 'Yes'," the meaning which the Greeks have chosen for this same fixed action pattern is that of "disapproval" as a "factual 'No'."

In these two instances we have decisive evidence of the way in which human groups, through the exercise of choices that are not genetically prescribed, create highly specific conventional behaviors. It is the existence of such conventional behaviors, in great profusion, in all human populations, that establishes, indubitably, the autonomy of culture. Moreover, it is these same conventional behaviors that make up the "ethnographic detail" which is the very subject matter of anthropology. And, it is precisely because these behaviors are the outcome of the human capacity to make choices, and not of any genetic process, that this "ethnographic detail" cannot possibly be explained by sociobiological theory.

Further, because cultural phenomena are particular alternatives, created by human agency in the course of history, it is always possible for these alternatives to be rapidly, and even radically, changed. The literature of anthropology abounds with reports of such changes which demonstrate that the process of cultural change, through the choice of new alternatives, is, in many instances, not connected in any significant way with the process of genetic evolution, or, for that matter, with human physiology.

This brings us to the crux of the argument, for it follows, given the exogenetic nature of very many cultural processes, that anthropology and human ethology could not conceivably be reduced to "the sociobiology of a single species" as Wilson (1975:547), in bemused mood, masquerading as a zoologist "from another planet," has chosen to suppose.

I conclude, therefore, that sociobiology, for all its pawings of the ground, presents no kind of major "antidisciplinary" threat to anthropology. Indeed, when it is realized that human evolutionary history has resulted in the emergence of exogenetic processes, and of a 2nd, or cultural, system of evolution, it becomes evident that socio-biological theory *per se,* being restricted to purely genetic processes, is categorically unfitted for the comprehensive scientific study of the evolution and behavior of that most uninsectlike of all creatures, for good or ill, the human animal.

NOTES

1. For a discussion of "the human activities of choosing" as "a special case of procedures that are fundamental to all living processes," see J. Z. Young (1978:12 *seq.*); for a discussion of the anthropological significance of human choice behavior, see Freeman (1978).
2. Cf. Itani and Nishimura (1973); Kawai (1965); Kawamura (1963), and van Lawick-Goodall (1973).
3. This conclusion bears on the long-standing anthropological problem of the origin of culture, and indicates that Kroeber's notion (1917: 176) of "the utter divergence between social and organic forces" is in error.
4. This conclusion has been reached by a number of evolutionary biol-ogists; e.g., Waddington, who refers to man having (1961a:70) "a second evolutionary system superimposed on top of the biological one, and functioning by means of a different system of information transmission."
5. Cf. Medawar, 1957:141; Dobzhansky, 159:81; Waddington, 1961b: 73; Huxley, 1966:16; Monod, 1969:16; Wynne-Edwards, 1972:64; Tinbergen, 1978:232.
6. Cf. Young, 1978:270, who refers to the "multi-parental inheritance of information."
7. The term *exosomatic* was first used by Alfred Lotka in 1945 (in a paper in the journal *Human Biology*) to describe the "entirely new" evolutionary "path" which had been followed by the human species in "more recent times." In identifying this same process (in a letter published in *The Times* (London), on May 30, 1910), E. Ray Lankester described the "transmission" of "tradition" in human societies as being *extra-corporeal.* In this present paper I have pre-ferred the term *exogenetic,* which has also been used by Medawar (cf. D. Wilson, 1971:29).

8. Cf. Trivers, 1976:v; "Natural selection has built us; and it is natural selection we must understand if we are to comprehend our own identities."
9. The phrase "with its own internal lightning blind" is from P. B. Shelley's "Letter to Maria Gisborne."

BIBLIOGRAPHY

Barash, D. P. 1977 *Sociobiology and Behavior.* New York: Elsevier.

Blakemore, C. 1977 *Mechanics of the Mind.* Cambridge: Cambridge University Press.

Clark, G., and S. Piggott 1970 *Prehistoric Societies.* Harmondsworth: Penguin Books.

Dawkins, R. 1976 *The Selfish Gene.* Oxford: Oxford University Press.

Dobzhansky, T. 1959 "Human Nature as a Product of Evolution." *In,* A. H. Maslow (ed.), *New Knowledge in Human Values.* New York: Harper and Brothers, pp. 75–85.

—— 1963 "Cultural Direction of Human Evolution—A Summation." *Human Biology,* 35:311–16.

Eibl-Eibesfeldt, I. 1975 *Ethology: The Biology of Behavior.* New York: Holt, Rinehart & Winston.

Flew, A. 1975 *Thinking About Thinking.* London: Fontana.

Freeman, D. 1974 "The Evolutionary Theories of Charles Darwin and Herbert Spencer." *Current Anthropology,* 15:211–37.

—— 1978 *A Precursory View of the Anthropology of Choice.* Department of Anthropology, Research School of Pacific Studies, The Australian National University.

Hawkes, J. 1976 *The Atlas of Early Man.* London: Macmillan.

Herodotos 1958 *The Histories.* Translated by H. Cary. London: Bohn.

Huxley, J. 1961 "The Humanist Frame." *In,* J. Huxley (ed.), *The Humanist Frame.* London: Allen and Unwin, pp. 13–48.

Isaac, G. L. 1976 "Stages of Cultural Elaboration in the Pleistocene: Possible Archaeological Indicators of the Development of Language Capabilities." *Annals, New York Academy of Sciences,* 280:275–88.

Itani, J., and Nishimura, A. 1973 "The Study of Infrahuman Culture in Japan." Symposia of the 4th International Congress of Primatology. Vol. 1: *Precultural Primate Behavior,* pp. 26–50.

Kawai, M. 1965 "Newly Acquired Pre-Cultural Behavior of the Natural Troop of Japanese Monkeys on Koshima Islet." *Primates,* 6:1–30.

Kawamura, S. 1963 "The Process of Sub-Culture Propagation Among Japanese Macaques." *In,* C. H. Southwick (ed.), *Primate Social Behavior.* Princeton: Van Nostrand, pp. 82–90.

Kroeber, A. L. 1917 "The Superorganic." *American Anthropologist,* 19:163–213.

Lankester, E. Ray 1910 "Heredity and Tradition." *The Times* (London) May 30, p. 9.

Lawick-Goodall, J. van 1973 "Cultural Elements in a Chimpanzee Community." Symposia of the 4th International Congress of Primatology. Vol. I: *Precultural Primate Behavior,* pp. 144–84.

Lotka, Alfred J. 1954 "The Law of Evolution as a Maximal Principle." *Human Biology, 17:*167–94.

Mayr, E. 1976a "The Evolution of Living Systems." *In, Evolution and the Diversity of Life: Selected Essays.* Cambridge: Belknap Press. Original publication 1964.

——— 1976b "Teleological and Teleonomic: A New Analysis." *In, Evolution and the Diversity of Life: Selected Essays.* Cambridge: Belknap Press. Original publication 1974.

——— 1976c "Behavior Programs and Evolutionary Strategies." *In, Evolution and the Diversity of Life: Selected Essays.* Cambridge: Belknap Press. Original publication 1974.

Medawar, P. B. 1957 "Tradition: The Evidence of Biology." *In, The Uniqueness of the Individual.* London: Methuen, pp. 134–42. Original publication 1953.

——— 1976 "Does Ethology Throw Any Light on Human Behaviour?" *In,* P. P. G. Bateson and R. A. Hinde (eds.), *Growing Points in Ethology.* Cambridge: Cambridge University Press, pp. 497–506.

Medawar, P. B. and J. S. 1977 *The Life Sciences: Current Ideas in Biology.* London: Wildwood House.

Monod, J. 1969 "From Biology to Ethics." *Occasional Papers of the Salk Institute of Biology,* Vol. 1.

Popper, K. R. 1974 *In,* F. Elders (ed.), *Reflexive Water: The Basic Concerns of Mankind.* London: Souvenir Press.

——— 1976 *Unended Quest: An Intellectual Autobiography.* London: Fontana.

——— 1977a "Contribution to Dialogue III." *In,* K. R. Popper and J. C. Eccles, *The Self and Its Brain.* Berlin: Springer International.

——— 1977b *Natural Selection and the Emergence of Mind.* The First Darwin Lecture, delivered at Darwin College, Cambridge, November 8, 1977.

Slobodkin, L. B. 1978 "Is History a Consequence of Evolution?" *Perspectives in Ethology, 3:*233–55.

Steiner, G. 1976 *After Babel.* Oxford: Oxford University Press.

Teng, E. L., P. Lee, K. Yang, and P. C. Chang 1976 "Handedness in Chinese Populations: Biological, Social and Pathological Factors." *Science, 193:*1146–50.

Tinbergen, N. 1978 "Use and Misuse in Evolutionary Perspective." *In,* W. Barlow (ed.), *More Talk of Alexander.* London: Gollancz, pp. 218–36.

Trivers, R. L. 1976 Foreword. *In,* R. Dawkins, *The Selfish Gene.* Oxford: Oxford University Press, pp. v–vii.

Waddington, C. H. 1961a "The Human Evolutionary System." *In,* M. Banton (ed.), *Darwinism and the Study of Society.* London: Tavistock, pp. 63–81.

——— 1961b "The Human Animal." *In,* J. Huxley (ed.), *The Humanist Frame.* London: Allen and Unwin, pp. 65–80.

Washburn, S. L. 1963 "The Study of Race." *American Anthropologist, 65:*521–31.

——— 1978 "The Evolution of Man." *Scientific American, 239:*146–54.

Wilson, David 1971 *The Science of Self.* London: Longman.

Wilson, E. O. 1971a *The Insect Societies.* Cambridge: Harvard University Press.

——— 1971b "Competitive and Aggressive Behavior." *In,* J. F. Eisenberg and W. S. Dillon (eds.), *Man and Beast: Comparative Social Behavior.* Washington: Smithsonian Institution Press, pp. 183–217.

——— 1975 *Sociobiology: The New Synthesis.* Cambridge: Harvard University Press.

——— 1976 "Sociobiology: A New Approach to Understanding the Basis of Human Nature." *New Scientist, 70:*342–44.

——— 1977 Foreword. *In,* D. P. Barash, *Sociobiology and Behavior.* New York: Elsevier, pp. xiii–xv.

——— 1977 "Biology and the Social Sciences." *Daedalus, 106*(4): 127–40.

——— 1978 *On Human Nature.* Cambridge: Harvard University Press.

Wynne-Edwards, V. C. 1972 "Ecology and the Evolution of Social Ethics." *In,* J. W. S. Pringle (ed.), *Biology and the Human Sciences.* Oxford: Clarendon Press, pp. 49–69.

Young, J. Z. 1978 *Programs of the Brain.* Oxford: Oxford University Press.

DAVID LAYZER

ON THE EVOLUTION OF INTELLIGENCE AND SOCIAL BEHAVIOR

I. *Sociobiology and Evolutionary Theory*

That human social behavior has a genetic basis no biologically literate person would nowadays deny. But human sociobiology, as expounded by Edward O. Wilson (1978) in his recent book *On
Human Nature*, goes beyond this simple claim. Wilson asserts that
human social behavior is constrained by genetic predispositions, including predispositions "toward learning some form of communal
aggression," toward "altruistic" behaviors that "serve the altruist's
closest relative and . . . decline steeply in frequency and intensity
as relationship becomes more distant," and toward religious belief.
The modern manifestations of these genetic predispositions are
"hypertrophic forms of original, simpler responses," analogous to
the elephant's tusk (a hypertrophied tooth) and the male elk's
antlers (hypertrophied cranial bones). Culture can modify the expression of genetic predispositions but it cannot permanently suppress them. "The genes hold culture on a leash. The leash is very
long, but inevitably values will be constrained in accordance with
their effects on the human gene pool." Nor will the insight into
these innate predispositions that sociobiology promises weaken their
hold: "I do not believe it [pure knowledge] can change the ground
rules of human behavior or alter the main course of history's predictable trajectory."

Granted that all behavior has a genetic basis, must one accept this view of human nature? "The question of interest," Wilson writes, "is no longer whether human social behavior is genetically determined; it is to what extent." But can one meaningfully ask *to what extent* before one understands *how?* Wilson's analogy between hypertrophied organs and social behaviors is not very helpful in this respect. Tusks and antlers evolved under selection pressures different from those that shaped the antecedent structures, and in each case the change of form reflected a change of function. Wilson asserts, however, that the genetic substrate of human social behavior has changed very little during the past 10,000 years and that modern behaviors fulfill essentially the same biological functions as their primitive antecedents. "Cultural hypertrophy" as envisioned by Wilson may perhaps be likened to the developmental hypertrophy of muscles through exercise. But this simile is not very illuminating.

Sociobiologists have approached the question of how genetic programs influence social behavior in two ways—one inductive, the other theoretical.

Wilson's discussion of aggression illustrates the inductive approach. Observations of aggressive behavior in diverse cultures suggest that particular environmental conditions regularly elicit particular aggressive behaviors. To explain such observations Wilson postulates that certain learning rules have been built into the human central nervous system in the course of its evolution: "to partition other people into friends and aliens, . . . to fear deeply the actions of strangers, and to solve conflict by aggression." He argues that natural selection would have favored the genetic encoding of such rules during the hunter-gatherer phase of human evolution, when the behaviors they specify would (he claims) have been biologically advantageous.

I think one could construct an equally persuasive argument in favor of genetically programmed learning rules for cooperative behavior, using observational data like those presented by Ashley Montagu in his recent book *The Nature of Human Aggression.* The fact that man's capacity to form large social groups based on mutual trust and division of labor vastly exceeds that of any other primate suggests that this capacity has a genetic basis, laid down during the hunter-gatherer stage of human evolution.

Learning rules that promote cooperative behavior would, however, conflict with the learning rules for aggressive behavior quoted
above. It is difficult—but not perhaps impossible—to understand
how two mutually contradictory sets of learning rules could have
been shaped simultaneously by natural selection.

Wilson and other sociobiologists avoid this difficulty by appealing
to a certain theory of altruistic behavior. According to this theory
a gene affecting behavior will be positively selected if and only if it
increases the reproductive potential of its carriers. A gene for altruistic behavior—behavior that benefits others at the altruist's expense
—may be selected if the behavior it specifies benefits other carriers
of the same gene. Wilson refers to such behavior as "hard-core"
altruism, to distinguish it from "soft-core" or, as Robert Trivers
(1971) called it, "reciprocal" altruism. Reciprocal altruism is ultimately selfish. Its "psychological vehicles," says Wilson, paraphras-
ing Trivers, are "lying, pretense, and deceit, including self-deceit."
Hard-core altruism strengthens family ties; soft-core altruism is the
glue that binds social units larger than the nuclear family. According to this theory human social organization rests on reciprocal
altruism and hence on "lying, pretense, and deceit, including self-
deceit." "The predicted result," Wilson writes, "is a mélange of
ambivalence, deceit, and guilt that continuously troubles the individual mind."

In the light of the theory just sketched, the learning rules underlying aggressive and cooperative behaviors are not fundamentally
antagonistic. They represent complementary strategies through
which the genes that specify these behaviors increase their representation in the gene pool.

Wilson's view of human nature relies on extrapolations from
observations of animal behavior interpreted in the light of a certain
theory. The theory serves two functions. It defines the class of relevant observations and it supplies the key to their interpretation. An
observer ignorant of the theory might fail to perceive the elements
of pretense, lying, and deceit underlying cooperative and altruistic
human behaviors. Noting that in most cultures adults behave in a
protective and nurturing way toward genetically unrelated infants,
such an observer might be tempted to infer that disinterested be-
nevolence has a genetic basis. Sociobiology teaches us that this can-

not be the case. Insofar as it is genetically encoded, behavior toward unrelated persons must be "ultimately selfish."

How secure is the theoretical basis for this conclusion? The applications of sociobiology reviewed by Wilson rest on two major premises. The first is that the evolutionary fate of any biological character is determined by its "inclusive fitness." As defined by Haldane (1932) and Hamilton (1964), the inclusive fitness of a character is its contribution to its possessor's reproductive potential augmented by suitably weighted contributions to the reproductive potentials of genetically related individuals. The doctrine of hard- and soft-core altruism springs from this premise, which implies that hard-wired behavior must initially have served to augment inclusive fitness and that ostensibly altruistic patterns of behavior must in reality be self-serving. The second premise is that the genetic determinants of social behavior act in a particular way. They do not specify behaviors in detail but make certain directions of cultural evolution more likely than others. Thus cultural evolution is not written on a blank sheet but on a "sheet at least lightly scrawled with certain tentative outlines" (Hamilton, 1975). In the same vein Wilson refers to the "predictable trajectory of human history." I shall argue that both premises are mistaken—that the evolutionary fate of a biological character need not be determined by its inclusive fitness and that an important class of human behaviors is not constrained in the manner suggested by the above metaphors.

1. The theory of inclusive fitness was developed within the framework of classical population genetics. But classical population genetics gives an inadequate account of the evolution of complex biological characters, for two reasons (Layzer, 1978a). It lacks effective techniques for describing the evolution of complex genetic systems. And it supplies no prescription for relating genic fitnesses —the theory's primitive variables—to the fitnesses of whole genotypes in specific environments. These defects of the classical theory are especially relevant to the evolution of social behaviors, which are specified by complex genetic systems and depend in a complicated way on interactions between the members of a population. I have elsewhere (Layzer, 1978a) described a formulation of population genetics that avoids these difficulties and I have used this theory to discuss the evolution of altruistic behavior (Layzer,

1978b). The main arguments are summarized in Section II below. They show that the fitness of an interactive trait may differ significantly from its inclusive fitness. An interactive trait may be positively selected even if it reduces its possessor's inclusive fitness. Group selection, which has usually been called upon to explain the rise and persistence of traits that diminish their possessors' inclusive fitness but promote the welfare of the group, need not be invoked. Natural selection operating at the individual level can promote genuinely cooperative behaviors—behaviors not based ultimately on pretense, lying, and deceit.

2. I shall argue that certain behaviors, though strongly influenced by genetic factors, are nevertheless not constrained in the manner envisioned by Wilson and other sociobiologists. Consider that quintessentially human behavior, learning to speak. Obviously it has a genetic basis. But what, precisely, do the genes specify? Not the grammar and vocabulary of any specific language; the ability to speak a *particular* language is not inherited. Chomsky has hypothesized that a tacit knowledge of certain deep-seated structural regularities common to all natural languages is genetically programmed. What are these regularities? What does it mean to have a tacit knowledge of them? And how did natural selection shape the genetic basis for this kind of knowledge? None of these questions has yet been answered. The chief argument offered in support of the hypothesis is indirect. Chomsky argues—in my opinion correctly—that a child cannot possibly acquire the linguistic knowledge implied by his linguistic behavior (at any developmental stage) by induction from experience. The syntactic rules implicit in a child's utterances are always underdetermined by his linguistic experience. The empiricist account, based on the dogma that experience alone shapes behavior, is therefore untenable. Rejecting it, Chomsky embraces a nativist account: A child needs so little linguistic experience to learn to speak because his mind is not a blank sheet but "a sheet at least lightly scrawled with certain tentative outlines." In this case the tentative outlines represent tacit knowledge of a hypothetical universal grammar.

Empiricism and nativism are classical epistemological stances, along with rationalism (the doctrine that there is a nonsensible reality whose properties we apprehend through pure reason). The rise of modern science, especially during the present century, has led to a new view of knowledge and the process by which it is acquired. According to this view knowledge is not induced from experience, discovered by pure reason, or shaped by perceptual and cognitive structures. Rather it is *constructed* through a process in which experience, reason, and *invention* all play essential roles. Scientists invent hypotheses, compare their logical consequences with experience, and, guided by the results of this comparison, construct new hypotheses. What is novel about this view is the realization that invention cannot be reduced to pure reason or induction, and is not constrained in the way that Kant believed by structural properties of the human perceptual-cognitive apparatus.

One might try to develop accounts of perception, cognition, and cognitive development along similar lines. Viewed in this way, perception would be a circular process in which the brain constructs a trial percept, matches it to relevant stimuli, reconstructs the percept, and so on, until an adequate match has been achieved. In cognition, "cognitive hypotheses" would be constructed and matched to relevant stimuli by an analogous process. In cognitive development, cognitive structures (Piaget's "schemata") play the role of hypotheses. In learning to speak, for example, a child would construct, test, and reconstruct a progressively more adequate repertoire of words and phrases and a progressively more adequate set of syntactic and phonological rules for their use.

Constructivist accounts of perception, cognition, and cognitive development have a common weakness: they do not explain how the key process of construction actually occurs. The analogy between scientific invention and perceptual or cognitive construction merely explains one obscurity by another. For example, Piaget refers to the process by which successive cognitive schemata are constructed as "equilibration," but the nature of the process remains obscure.*

This gap in constructivist accounts of perception, cognition, and cognitive development has a counterpart in the prevailing theory

* There is a curious analogy between Piaget's theory that cognitive structures evolve by successive "equilibrations" through a series of qualitatively distinct forms and Prigogine's theory of successive quasi-equilibrium states in certain dissipative systems. I do not know whether Piaget attaches any significance to this analogy.

of biological evolution, according to which biological adaptations are shaped by natural selection from the raw material provided by genetic variation. Although few biologists doubt that this account is basically correct, it seems to be seriously incomplete. Attempts to construct mathematical models of the evolutionary process based on realistic estimates of mutation and recombination rates *and on simple assumptions about genetic variation* have invariably predicted evolutionary stagnation.

If the theory's basic biological premises are correct, its inability to describe the evolution of complex adaptations must result from oversimplified assumptions about genetic variation. Genetic variation must be blind with respect to its phenotypic consequences: that much is established beyond reasonable doubt. But blind variation need not be "random" in the sense of "unstructured" (Mayr, 1963; Dobzhansky, 1970). I have proposed (Layzer, 1978a, 1980) that genetic variation (mutation and recombination) is controlled by a genetic system (the "β-system") that evolves, just as the genetic system that specifies the development of the phenotype (the "a-system"), by blind variation and natural selection. The β-system serves to direct genetic variation into potentially adaptive channels and to suppress or buffer potentially disruptive variations. I have argued that the β-system specifies a progressively more elaborate evolutionary strategy of *hierarchic construction.*

The thesis that biological evolution proceeds by hierarchic construction was put forward long ago by I. I. Schmalhausen (1949) on the basis of comparative embryology. The evolution of genetic material through gene duplication, differentiation, and integration is likewise known to proceed by hierarchic construction. In the paper cited earlier (Layzer, 1980) I argued that the evolutionary strategy of hierarchic construction is itself the product of an evolutionary process.

In the present essay I shall argue that there is a deep-seated analogy between biological evolution and certain forms of human behavior. Biological evolution is creative in the sense that it generates novel solutions (adaptations) to unforeseen problems posed by the environment. I shall argue, following Mayr (1974), that certain human behaviors are also creative in this sense and that the capacity for creative behavior is itself a biological adaptation. This notion will enable us to reconcile the fact that human behavior has a genetic basis with the thesis that certain kinds of human behavior are not constrained in the manner described by contemporary sociobiologists.

The plan of the essay is as follows. Section II deals with theories of altruism. Section III treats the process of hierarchic construction in evolutionary and developmental contexts. Section IV treats creative behavior and theories of its development. Section V discusses the role of logical structures in cognitive development. Section VI discusses language acquisition from the perspective of hierarchic construction. Finally, Section VII advances a new hypothesis concerning the relation between intelligence and society.

II. *Theories of Altruism*

Altruistic behavior, by definition, increases the reproductive potential of others while reducing the altruist's own. At first sight one might suppose that natural selection would tend to eliminate any genetic predisposition to behave altruistically, but this conclusion does not allow for the fact that the beneficiary of an altruistic act may carry copies of the genes that specify the act. Haldane (1932) cited the example of a person who risks drowning to rescue a drowning relative. The genes that specify the impulse to such an act will spread in a population if the (statistical) expectation of reproductive benefit *for these genes* exceeds the expectation of reproductive cost. Thus the level of acceptable risk rises with the degree of genetic relationship between the altruist and the beneficiary. The "inclusive fitness" of a behavioral trait includes suitably weighted contributions from genetically related beneficiaries of the behavior.

These ideas have been applied with conspicuous success to the social insects, whose elaborate social behaviors have been shaped by natural selection. As first pointed out by W. D. Hamilton, a crucial factor in this process is the trait of haplodiploidy. Unfertilized eggs develop into haploid males, fertilized eggs into diploid females. Hence the degree of genetic relationship between sisters (i.e., the average proportion of shared genes) is $3/4$, while the degree of genetic relationship between mother and daughter is ½. Because "sisters [are] more closely related to each other than mothers are to daughters . . . females may derive genetic profit from becoming a sterile cast specialized for the rearing of sisters. Sterile casts engaged in rearing siblings are the essential feature of social organization in the insects" (Wilson, 1978:12).

Like the social insects, human beings form large social groups based on cooperation and division of labor. But such groups may consist mainly of genetically unrelated individuals, so the genetic basis for human social organization must be completely different from that of social organization among wasps, bees, and ants. As mentioned in Section I, Wilson and other sociobiologists have concluded from the theory of inclusive fitness sketched above that human social behavior is based on soft-core or reciprocal altruism, which in turn rests on deception and self-deception. I have argued elsewhere (Layzer, 1978b) that this conclusion is not justified because "inclusive fitness" is too specialized a concept. In the following paragraphs I shall try to make this conclusion plausible by qualitative reasoning. A detailed mathematical argument may be found in the reference just cited.

The fitness of a genetic trait (i.e., its multiplicative contribution to the reproductive potential of a typical bearer of the trait) depends on how it is expressed in the phenotype. If two genetically distinct traits are phenotypically indistinguishable in a certain range of environments they also have the same fitness in this range. For interactive traits the situation is somewhat more complicated. Let us consider a simple example, warning calls in birds. This behavior has two aspects: emitting a warning call when a predator is sighted, and heeding the warning call emitted by another bird. The first aspect entails a certain risk. Taken by itself it tends to diminish fitness. The second aspect confers a benefit; it increases fitness. The behavior as a whole has positive fitness if the expectation of benefit exceeds the risk. Evidently the expectation of benefit increases and the risk decreases as the number of birds participating in the behavior increases. If N birds give and heed warning calls each bird will receive warning calls $N - 1$ times as frequently as he gives them. It doesn't matter whether the genetic systems that specify the behavior are identical in different birds, provided the warning calls are mutually intelligible. And natural selection will favor the evolution of mutual intelligibility. Thus in a large flock the fitness of warning-call behavior may greatly exceed its "inclusive fitness," calculated in the conventional way.

The phenotypic fitness of a trait reduces to its inclusive fitness if the individuals carrying that trait or a functionally equivalent one are confined to a single lineage. This would be the case for a trait that resulted from a highly improbable sequence of individually improbable mutation and recombination events. The spread of such a trait in a population would be determined, apart from statistical fluctuations, by its inclusive fitness. But do complex adaptations in fact arise through highly improbable sequences of individually unlikely mutation and recombination events confined to a single lineage? According to the theory cited earlier (Layzer, 1978a, 1980), complex adaptations usually arise simultaneously *and in many variant forms* in several distinct lineages in a population. In these circumstances the phenotypic fitness of a complex interactive trait may substantially exceed its inclusive fitness. As in the example of warning calls, natural selection promotes functional convergence of interactive behaviors specified by distinct but homologous genetic systems in a population.

How stable are social behaviors against genetic variations that increase one's own inclusive fitness while diminishing that of other members of the population? Antisocial behaviors do of course occur in human social groups. But because social behaviors as a whole make such an important contribution to fitness, we may expect genetic predispositions to conform and to exact social conformity from others to be strongly selected. Such predispositions reduce both the incidence and the inclusive fitness of antisocial behaviors. Waddington (1961) has argued that the human tendency to construct and conform to ethical codes can be understood in this way— as an adaptation that reinforces fitness-enhancing social behaviors.

To sum up, the evolutionary fate of an interactive trait depends in general not on its inclusive fitness but on its phenotypic fitness, which may include substantial contributions from genetically unrelated individuals. In man language makes possible interactive behaviors that knit together social groups much larger than the nuclear family. If one accepts the theoretical framework referred to earlier (Layzer 1978a, 1980), it is reasonable to suppose that cooperative behaviors and a set of emotional responses that reinforce them would have been strongly selected during the earliest stages of human evolution.

The present theoretical framework allows us to see the relationship between aggressive and cooperative behaviors in a new light. There are two competing strategies for increasing the fitness of interactive social traits. One is to strengthen interactions within the group, the other is to make the group larger. Most species have limited resources for implementing the second strategy. But in man language makes possible the formation of social communities that cut across familial, tribal, and racial divisions. Yet language and other aspects of culture, such as religion, can also serve to increase the isolation and mutual hostility of social groups. Natural selection has shaped the genetic capacity to choose between these two strategies but has left the choice itself free.

III. *Hierarchic Construction*

The development of an individual organism occurs through interaction between the genome and the environment. Development is in part a process that translates information encoded in the genes and in part a process that creates information. The development of tissues and organs, for example, is largely a matter of translating genetic information, and so is the development of many behaviors. Certain human behaviors, however, not only translate but generate information. I shall refer to these as *creative* behaviors. I shall argue that there is a deep analogy between the development of creative behaviors and the evolution of adaptations, in that both occur through a process of hierarchic construction governed by a certain "logic."

A developing or evolving biological structure must continue to function as it grows more complex. That is why differentiation of structure and function must always be accompanied by integration. The genome as a whole must always contain an integrated program for development, and each subprogram must continue to fulfill its essential functions as it grows more complex through internal differentiation. Analogously, the human infant's first utterances are fully functional. They are not sentence-fragments but whole sentences expressed by single words—"holophrases." Each stage of linguistic development brings new differentiation of sound and meaning within a conservative integrated structure.

Hierarchic construction is at the same time a conservative and an innovative process. Structure and function become more complex as innovations are incorporated and stabilized. Stabilization and innovation play complementary and equally essential roles. Without innovation there would be stagnation; without stabilization, chaos.

The basic mechanisms of hierarchic construction in biological contexts are (a) duplication (replication) of existing structures, (b) variation (resulting in the differentiation of duplicated structures), (c) integration, and (d) selection. These mechanisms are necessary, but they are not sufficient to ensure that hierarchic construction will actually take place. The following example will show why.

Consider a collection of rigid rods of differing lengths from which a framework is constructed in accordance with the following stipulations. (a) Any rod or group of rods may be *duplicated*. (b) Any rod may be *varied* in length in any way consistent with the mechanical constraints imposed by the structure. (c) Rods or groups of rods may be *linked* by joints that allow the linked component to rotate freely about one another.

If we begin with a collection of unlinked rods and allow these three processes to occur at random but at specified average rates, they will soon produce a collection of more or less complex structures. (d) From among these structures we continually *select* "successful" ones according to fixed but inscrutable criteria. (Natural selection uses criteria that are inscrutable at the genetic level.)

A computer program for simulating the construction of frameworks in accordance with these stipulations would need subroutines for varying the lengths of rods, joining them together, and duplicating existing structures. It would also need subroutines for generating random variations and random combinations of elements. Finally, it would need subroutines for evaluating structures according to the "fixed but inscrutable" selection criteria. But a program containing *only* these subroutines could not carry out an extended hierarchic construction. At first more and more complex structures would be built up, but eventually progress would cease. The more elaborate a structure, the larger the number of potentially disruptive variations and the smaller the proportion of potentially constructive ones. When a certain level of complexity had been reached, random variations would disrupt old structures at the same rate as it created new structures.

To preserve "successful" structures that have already been built up one needs an additional subroutine to regulate the *amplitude* of the random variations. During the initial stages of construction this subroutine would instruct the computer to search for "good" frameworks composed of a small number of rods. Since the computer memory initially contains no information about what constitutes a "good" size or shape, the amplitude of the random variations should be large to begin with and should taper off as a "fitness peak" is approached. Once the range of good "primary" structures had been mapped out the regulatory subroutine would direct the computer to start linking together these structures to form "secondary" structures. At this stage variations of the primary structures would need to be subordinated to constraints imposed by the secondary structures in which they are being incorporated. The evolution of tertiary and higher-order structures would proceed analogously.

This example suggests that *regulation of variability* is essential for hierarchic construction. In the absence of effective regulation the process of hierarchic construction would become increasingly slow and ineffectual, and at some moderate level of complexity it would bog down completely. That stabilization is a prerequisite for extended hierarchic construction has previously been recognized by several evolutionary biologists, including I. I. Schmalhausen (1949) (who invoked "stabilizing selection"), L. B. Slobodkin (1964) (who talked about "conservation of homeostasis"), and G. L. Stebbins (1969) (who suggested that a principle of "conservation of organization" needed to be added to the list of basic evolutionary postulates).

In the papers cited earlier (Layzer, 1978a, 1980) I have argued that the regulation of genetic variation is accomplished by a genetic system whose elements regulate mutation and recombination rates. This genetic system itself evolved through random genetic variation and natural selection. It serves to direct genetic variation into potentially adaptive channels and to suppress or buffer potentially disruptive variation. In short, it expedites the evolution of adaptations and then stabilizes the evolved structures and functions. Since this genetic system is itself an adaptation it expedites its own evolution. Thus hierarchic construction, including the stabilization of deep-lying hierarchic levels, is not an independent biological principle but a consequence—or, more precisely, an expression—of the regulation of genetic variability by a genetic system that has itself arisen through random genetic variation and natural selection.

As discussed in Section I, biological evolution furnishes a paradigm for constructive processes in other biological contexts. Abstracting from the specific biological mechanisms involved in the construction of genetic material and the selection of phenotypes, we may formulate three general theses:

1. In all biological contexts the functional organization of complex structures (including mental structures such as percepts and meanings) is hierarchic.

2. All such hierarchic structures are generated (in the course of evolution, development, or experience) by a process of hierarchic construction in which new units of organization arise through simultaneous differentiation and integration of existing units.

3. There is a universal strategy of hierarchic construction. Its essential components are: (a) duplication of existing structures, (b) random variation of existing structures, (c) linking (or integration) of existing structures, (d) regulation of variability, and (e) selection according to fixed criteria that are inscrutable at the level of variation.

IV. *Creative Behavior*

All behavior has a genetic basis. But not all behaviors are genetically conditioned in the same way (Mayr, 1974). At one extreme are behaviors of varying degrees of complexity whose development seems to be programmed in much the same way as that of physical organs. Experience is necessary for their development but the adult form of the organ or behavior is insensitive to the quality or quantity of experience, provided these lie within certain broad limits (the "reaction norm"). The behaviors of social insects are of this kind. Despite their complexity they are as predictable as reflexes.

Besides genetically programmed behaviors there are learned behaviors. The capacity to learn is of course itself a biological adaptation, enabling an animal to cope in a more flexible way with a wider range of environmental challenges. However, the capacity to learn has a biological cost. To the extent that a behavior is learned it cannot be genetically programmed. And while it is being learned it is less useful than the same behavior would be if it were genetically programmed. Whether the advantage of flexibility outweighs the disadvantage of not having instant access to finely adapted but stereotyped behaviors will depend on the nature and range of environmental challenges that an animal must face. Conversely an animal's capacity for learning determines the kind and range of environments that it can select.

The category of learned behaviors is itself a continuum. At one extreme the effects of learning may be analogous to the hypertrophy of a muscle through exercise. Or learning may simply direct development into one of several genetically predetermined channels. Further along the continuum, learning may elaborate the gene script as an actor uses gestures and vocal inflections to bring to life the script of a play or as a violinist "ornaments" a melody in Baroque music. As discussed in Section I, this is the kind of learned behavior that Wilson and sociobiologists seem to have in mind when they talk about "genetic predispositions."

Finally, at the other end of the continuum, beyond behaviors that merely elaborate genetically programmed patterns, there exist —or so I assert—creative behaviors, behaviors that are not prefigured in an animal's genetic program. The central thesis in Piaget's monumental contribution to developmental psychology is that the structures mediating creative behaviors (including cognitive behavior) are built up in the course of development by a dialectical interaction between the organism and the environment. This construction starts, of course, from genetically programmed structures elaborated during prenatal development, but its outcome is not, Piaget insists, prefigured in the genes. I shall not attempt to summarize the evidence and the arguments that support this thesis. Because of its central position in Piaget's account of cognitive development it may be said to be supported by the whole of the vast yet coherent collection of experimental and theoretical investigations into cognitive development that he and his collaborators have carried out during a research career that is now well into its second half-century.

And yet the thesis that there exist genuinely creative behaviors (in the sense just defined) presents a fundamental difficulty. Piaget recognized and tried to solve it in *The Origins of Intelligence in Children* (1936). Since then he has returned to it again and again, most recently in *Behavior and Evolution* (1976). Piaget recognized at the outset that the development of cognitive structures both parallels and continues the evolution of organic adaptations. Cognitive structures are organs that mediate the flow of information into an organism as the organs that serve metabolism mediate the flow of energy. And the process of evolution is identical in its formal aspects with the process of cognitive development. Both can be characterized as processes of hierarchic construction (although Piaget does not explicitly use this term). Evolutionary biologists such as Dobzhansky (1970) and Mayr (1963) have stressed that evolution is a creative process. Although it proceeds in accordance with deterministic microscopic laws, its outcomes are nevertheless unique and unpredictable. They are in no sense implicit in the physico-chemical laws that govern genetic material. Similarly, although cognitive development proceeds from genetically prescribed structures in accordance with deterministic microscopic laws, it generates structures that are unique and unpredictable, and in no sense implicit in either the initial structures or the laws that govern the physico-chemical processes through which the initial structures are transformed. But *how?*

Because the mechanisms underlying biological evolution are more accessible than those underlying cognitive development, Piaget has devoted considerable attention to this question in its biological context. In particular he has argued that the prevailing theory of evolution does not and cannot account for the evolution of complex adaptations, especially behavioral adaptations. Piaget argues, in effect, that undirected random genetic variations would inevitably give rise to evolutionary stagnation:

For example, inasmuch as the three prerequisites of an adapted bird's nest are its solidity, the protection it affords against predators, and its capacity to maintain a certain temperature, are we going to assume that in the case of the common swallow, say, there were once pairs which, under the influence of one of a range of mutations, built frail nests in perilous locations and left them full of holes for drafts to blow through, and that such behavior continued until the occurrence of happier mutations made possible the selection of more skillful individuals and the elimination of the inept? Obviously not. [Piaget (1978:148)]

To meet this difficulty Piaget advances the bold hypothesis that information about the outcomes of genetically programmed behaviors is somehow fed back to the germ cells where it is utilized to regulate genetic variation. In this way behavior becomes the "motor of evolution."

This hypothesis must, I think, be rejected. Not only does Piaget fail to suggest any mechanism by which *specific* information about the consequences of behaviors could be incorporated into the gene script, but even if such a mechanism existed it would not be able to do what Piaget would like it to do. It could, perhaps, bring about improvements and refinements of behaviors and their supporting physiological structures. But it cannot account for the active exploratory behavior that initiates all innovation.

The hypothesis of selection-regulated genetic variability does account for exploration and innovation. Natural selection promotes variability of those aspects of the genetic program that have not yet been optimized, because adaptive innovations evolve in lineages that carry genes specifying such variability. At the same time the theory accounts for the suppression of potentially harmful genetic variation. Thus it remedies those weaknesses in the prevailing theory that Piaget has criticized with so much skill and insight, but in a way that respects the "central dogma" of genetics. Moreover, the genetic mechanisms postulated by the theory (genes and genetic systems that regulate mutation and recombination rates) are known to exist (see Layzer, 1980, for examples).

Analogous to the prevailing theory of organic evolution are what Piaget (in *The Origins of Intelligence in Children*) calls "groping" or "trial-and-error" theories of development, in which behaviors are emitted at random and then selected (or reinforced) according to their outcomes. Complex behaviors are shaped by selection from random combinations of simpler behaviors. Piaget acknowledges that such accounts of development are true as far as they go. But he points out that groping behavior is always to some extent systematic. It is organized and directed by behavioral and cognitive structures already present. Nonsystematic groping occurs only during the preliminary exploratory phases of development that pave the way for increasingly systematic groping. But even nonsystematic groping is never *purely* random: it is always at least vaguely directed. Thus (he concludes) instead of distinguishing two qualitatively different kinds of groping we should talk about a continuum of progressively more organized and directed behaviors.

Perhaps the most distinctive aspect of Piaget's account of cognitive development is its treatment of the interaction between cognitive (including sensorimotor) structures and external reality, and the way in which this interaction leads to the development of progressively more adequate structures. The key notions in this account are *assimilation* and *accommodation*. Piaget calls assimilation the "basic fact of psychic [i.e., mental] life." Assimilation is the incorporation of perceptual stimuli into already existing structures (perceptual, cognitive, sensorimotor, etc.). These structures may be unaffected or they may be altered. Following James Mark Baldwin, Piaget refers to such an alteration as "accommodation."

Piaget acknowledges that the concept of assimilation needs to be clarified. It is, he says, "a concept . . . so fraught with meaning that it might seem equivocal." And he remarks that "invoking the concept of assimilation does not constitute an explanation of assimilation itself." The same may be said of "equilibration" or "autoregulation." These terms refer to the processes of constructing progressively more adequate quasi-equilibrium (i.e., relatively stable) structures in response to the disequilibrating influence of new environmental challenges and opportunities. As in evolution, the elaboration of progressively more adequate structures creates qualitatively new opportunities for interaction with the environment. These result in new challenges to the existing structures. That is, they stimulate the process of accommodation. This description seems to fit the observations very well. But it does not, as Piaget well recognizes, *explain* the evolutionary tendencies to build up ever more complex structures that interact ever more widely with the environment.

The theory of hierarchic construction sketched in the last section may provide a framework for an account of mental development that would fill these gaps in Piaget's theory. In this account "assimilation," "accommodation," and "equilibration" do not figure as irreducible concepts. They are replaced by functional descriptions of certain psychophysiological processes analogous to genetic variation, regulation of genetic variation, and selection of phenotypes. "Accommodation" corresponds to what I have called "hierarchic construction," the progressive elaboration of a continuously functioning structure through differentiation accompanied by integration. In the evolutionary context, hierarchic construction depends on the four basic processes mentioned earlier: duplication of existing genetic structures, variation, selection, and regulation of variation (through variation and selection of β-genes). I hypothesize that each of these processes has its functional analog in the development of perceptual, cognitive, sensorimotor, etc., structures. Piaget has already stressed (in *Origins*) that repetition and variation of structured activities are central to the development process. Repetition and variation are in fact the essential components of groping. The psychophysiological analog of natural selection must consist in processes that evaluate trial structures (presumably by matching them with appropriately processed stimuli or with previous constructs) and re-create them at a rate determined by the outcome of the evaluation. (The analogy suggests that the physiological correlates of skills, habits, memories, etc., are structures that are continually replicated.) Finally, the regulation of genetic variability by natural selection corresponds to the regulation of groping and, more generally, the shaping of effective strategies for constructing progressively more adequate structures and preserving the successful outcomes (through suppression of harmful variation).

This schematic account, which is meant to apply to all sorts of constructive developmental processes from perception to abstract thought and the development of cognitive schemata, raises both formal and substantive questions. Does selection-regulated variability actually do what I have claimed it does? Does it shape a strategy of hierarchic construction? Does it direct variability into potentially adaptive channels and suppress nonadaptive variability? These formal questions, though addressed in the papers cited earlier, have not yet been answered definitively. The substantive questions concern the mechanisms that serve to implement this strategy in specific contexts. What are the neurological mechanisms of hierarchic construction? What is the mechanism of "matching" through which such structures are "evaluated"? What mechanisms regulate the variability of neurological structures, and how did such mechanisms evolve? Needless to say, none of these questions has yet been answered. There is not a shred of direct evidence to support the hypothesis that hierarchic construction, directed by selection-regulated variation, underlies the development of perceptual and cognitive structures. The hypothesis does, however, have certain behavioral implications, which are more accessible to experiment and observation.

As in Piaget's account, development occurs through the progressive differentiation and integration of functional schemata. Hierarchic construction is initiated by the random variation of structural and (above all) regulatory elements that mediate the processes of differentiation and integration. The amplitude and frequency of these variations are regulated by "selection, i.e., by the benefits that accrue to the organism in consequence of the schemata under construction. Selection of the range and frequency of groping directs the hierarchic construction of schemata that cope with environmental challenges and exploit environmental opportunities. In the absence of challenges and opportunities construction must cease, at least temporarily, because in these circumstances the groping processes that instigate and sustain it are unprofitable and hence will be counter-selected. But what constitutes the "environment"? For culture-specific behaviors and their underlying schemata culture itself is the dominant environmental component. The cultural environment determines the kinds of questions it is profitable or even possible to ask and the kinds of knowledge it is useful to have. It reinforces the development of schemata that mediate its own internalization. Thus children reared in closely similar cultural environments will develop closely similar perceptual/cognitive schemata in a similar order and at similar ages.

There is no reason to expect that the same will be true of children reared in substantially different cultural environments. Cognitive development across cultures may perhaps turn out to resemble embryonic development across species. In the earliest stages of development the embryos of a fish, a pig, a chimpanzee, and a human being are virtually indistinguishable. As development proceeds the anatomical features that differentiate progressively narrower taxa emerge. I suggest that the development of perceptual/cognitive structures in different cultures may exhibit a similar branching. The development of the earliest schemata, which mediate perinatal behavior, is probably determined by genetic programs that do not vary systematically from culture to culture. As development proceeds, culture assumes an increasingly important role. Even at the sensorimotor level cultural differences are likely to be important. Finally, in the development of linguistic/cognitive schemata culture may play the dominant role.

The analogy between embryonic development across species and cognitive development across cultures must not, of course, be pushed too far. None of us is a prisoner of his cultural environment. Cultural boundaries are not insurmountable barriers, but to cross them one must build up new schemata. Indian logic is as incomprehensible to a Western logician who has not made a special study of it as Indian music is to someone whose musical training has been confined to the Western classical tradition.

Piaget aptly refers to the development of cognitive schemata as a *construction* of reality. According to Piaget, this construction occurs through a dialectical interaction between the child's cognitive structures on the one hand and external reality on the other. In the course of development the cognitive structures become progressively more adequate and, at the same time, progressively more abstract —that is, progressively more remote from immediate experience. The present account of cognitive development fully supports these generalizations. The remoteness of a cognitive schema from experience is measured by the number of hierarchic levels that separate it from the simplest perceptual schemata.

An account of cognitive development based on the theory of hierarchic construction would coincide with Piaget's account through the stage of sensorimotor intelligence. Thereafter it would begin to diverge. The two accounts would differ most conspicuously in their approach to "reflective" or creative intelligence—mental activity not dominated by immediate practical goals. This is not the place to explore these differences in detail, but because they are relevant to one of the central topics of this essay (the biological nature of creative behavior) I shall indicate briefly how and why they arise.

Piaget's account of cognitive development beyond the sensorimotor stage is dominated by the notion that there is a logic inherent in the coordination of actions. The child first constructs schemata (conservation, seriation, etc.) that embody this logic at the practical level. He then reconstructs them at the level of internal representation. According to Piaget, the logic inherent in the coordination of actions is a fragment of formal deductive logic and naïve set theory. Post-sensorimotor intelligence is thus characterized by the growth of logico-mathematical structures.

I suggest that such a description of cognitive development is valid only in certain cultural contexts, and that even in these contexts it may be seriously misleading.

To begin with, cognitive development already has a logic—the logic of hierarchic construction. Continued hierarchic construction automatically generates schemata of increasing abstractness and adequacy. Consider, for example, Piaget's conservation schemata. At age 7–8 Piaget's subjects come to understand that the quantity of liquid in a beaker remains the same when the liquid is poured into another beaker of a different shape. Piaget describes this advance in terms of (concrete) "reversible operations." But I think an equally satisfactory description can be given in terms of hierarchic construction: The notion of quantity of liquid in a beaker is generated by simultaneous processes of differentiation and integration. The quantity of liquid is differentiated from its height in a particular beaker. At the same time, the notion "quantity of liquid" serves to integrate into a single schema (a) the height and cross-sectional area of the liquid, and (b) the aggregate of liquid-filled beakers whose contents are interchangeable. Pre-conservers acknowledge that liquid is not lost or gained when the contents of one beaker are poured into another but deny that the short, fat beaker contains as much as the tall, thin one. The notion of conserved quantity integrates the set of possible shapes of a given "lump" of liquid.

The child's construction of time as a dimension of uniform mo-

tion is somewhat analogous. So long as we focus attention on individual examples of uniform motion, distance traversed serves to
measure both elapsed time and speed. The three notions are differentiated only when we construct an integrated description of two
or more uniform motions going on at the same time. This is the
way Aristotle describes the construction of time and speed in his
Physics, and it is also, according to Piaget's observations, the way
children construct the notion of time *in such contexts.*

Deductive logic is itself a product of hierarchic construction. But
it is not an inevitable product; it does not arise automatically when
a certain level of abstraction has been reached. Logic is not, as Piaget
describes it, the "mirror of thought" or the "axiomatics of reason."
Rather, it is a product of human intelligence—an invention whose
ramifications (which include mathematics, the natural sciences, and
science-based technology) have come to dominate Western culture.
Not only science and technology but also the humanities, rooted
as they are in classical Greek thought, foster respect for deductive
logic. We are taught very early to recognize, construct, and value
logical arguments. But recognizing or constructing a logical argument is not at all the same thing as thinking logically. Is Piaget
correct in asserting that fully developed human thought is logical?

V. *Deductive Logic and Bio-logic*

Within the framework of the present theory, the last question can
be given a more precise form: How is deductive logic related to
the logic of hierarchic construction? At first sight the two seem
closely related. Deductive logic assigns truth-values to formal sentences; natural selection (or its behavioral analog) assigns fitnesses
(or their behavioral analog) to biological structures. Just as deductive logic concerns itself with the formal structure of sentences
but not with their meanings or with the grounds for assigning
truth-values to them, so bio-logic (the logic of hierarchic construction) concerns itself with the formal structure of hierarchies but
not with their biological content or with the grounds for assigning
fitnesses to them. We shall see, however, that truth and fitness play
different roles in their respective contexts. This difference will make

it impossible to consider deductive logic as a special case of bio-logic. Thus if we accept the hypothesis that cognition and cognitive development utilize the logic of hierarchic construction, we shall have to reject Piaget's view of deductive logic as the "axiomatics of reason."

Deductive logic has its own recipes for hierarchic construction. Let $A, B, C, \ldots$ denote "atomic" sentences. The unary operator "$-$" may be applied to any sentence to yield a new sentence. Applied to A it gives the sentence $-A$ ("not$-A$"). Applied to the sentence $-A$ it gives back A: $-(-A) = A$. The binary operator "&" links any two sentences to form a new sentence. Applied to the atomic sentences A, B it gives the sentence $A\&B$, which we postulate is the same as $B\&A$. With the help of these two operators we can now construct sentences of arbitrary complexity.

Atomic sentences may be differentiated into subject and predicate. If F represents a predicate and x a subject, we may write $A = F(x)$. Predicates with two or more subjects express relations. Thus in the sentence $E(x,y)$, E could stand for the relation "equals." Then $E(x,y)$ would stand for "$x = y$."

Finally, predicates may occur as subjects of other (second-order) predicates. This if Φ is a second-order predicate, $\Phi[F(x)]$ is a sentence.

The preceding rules enable one to construct sentences complex enough to express the most intricate logical relations involving finite collections of individuals (subjects). Logical connectives other than "$-$" and "&" can all be expressed in terms of these two. For example, "$A\lor B$" ("A or B") is simply an abbreviation of $-(-A\&-B)$.

But we have not yet constructed a deductive logic. To the "syntactic" rules just described we must add rules for assigning truth-values to sentences. The simplest such rules are the following: (i) To every atomic sentence A assign the truth-value $t(A)$, where $t(A)$ is either 1 or 0. (ii) To $-A$ assign the truth-value $t(-A) = 1 -t(A)$. (iii) To $A\&B$ assign the truth-value $t(A) \times t(B)$. These three rules suffice. With their help we can calculate the truth-value of any complex sentence, given the truth-values of its atomic constituents.

We have now constructed a deductive logic—one that is too

weak for mathematics (because it is confined to finite collections of individuals) but powerful enough to express the most intricate logical relations among the members of any finite collection of individuals and predicates.

We are now in a position to investigate the analogy between deductive-logical structures and biological structures. Let $a_1a_2a_3 \ldots$ denote (a) a string of primitive logical signs $[-, \&, (,), $ predicate-symbols, individual-symbols], (b) a sequence of primitive DNA segments (codons). If the string of logical symbols has been constructed according to the rules given above, it represents a sentence capable of being assigned a truth-value. Analogously, if the string of codons has been constructed in the course of biological evolution, it specifies the development of an organism and can be assigned a certain fitness. But the way in which a truth-value is assigned to a complex sentence is entirely different from the way in which fitness is assigned to a genome: *The truth-value of a complex sentence derives from the truth-values of its atomic constituents; the fitness of genetic subroutines derives from the fitness of the genome as a whole, and the fitness of sub-subroutines derives from the fitness of subroutines.* This difference is fundamental. The fitness function $f(a_1a_2a_3 \ldots)$ not only determines the fitnesses that can be assigned to subroutines and sub-subroutines. It also determines the functional organization of the genome. That is, it determines which aggregates of genetic elements define subroutines, which subaggregates defined sub-subroutines, and so on. For example, the string $a_1a_2a_3$ defines a subroutine if and only if the equality $f(a_1a_2a_3a_4a_5 \ldots) = f(a_1a_2a_3)f(a_4a_5 \ldots)$ holds to a good approximation, where the "partial fitnesses" on the right side of this equation are constructed according to procedures discussed in detail in a reference cited earlier (Layzer, 1978a).

These considerations show that bio-logic—the strategy of hierarchic construction based on selection-regulated variability—can never mimic deductive logic. "Logical thought," according to the present theory, is thought that uses the previously constructed schema of deductive logic to recognize and construct logical arguments. But the *processes* of recognition and construction are biological rather than logical. Deductive logic supplies criteria that "logical thought" uses to assess the adequacy of its constructs, but it has nothing to do with the strategy of construction.

VI. *Language Acquisition*

The general thesis that language acquisition is a process of hierarchic construction, occurring through the simultaneous differentiation and integration of sound and meaning, is supported by recent systematic studies in psycholinguistics. Thus Roger Brown in his masterly study of the early stages of language acquisition, *A First Language,* characterizes early linguistic development in the following terms:

> This development is always of the same two kinds. An increase in the number of relations expressed by: 1. concatenating, serially, more relations and omitting redundant terms; 2. unfolding of one term in a relation so that the term becomes itself a relation. In these data as a whole, as also in Stage II and in Brown and Hanlon's (1970) results with tag questions, there is evidence for what I have, not yet very seriously, called a law of cumulative complexity in language development. It is important to realize that as utterances get longer, and MLU (mean length of utterance) increases, some sort of increase in complexity is bound to occur, but there is no a priori reason why the increase should take just the forms it does and, in particular, that these forms should be the same for all children studied, whatever the language in question. [Brown (1973:64–65)]

The two kinds of development Brown describes correspond precisely to what I have called integration and differentiation, and his "law of cumulative complexity" is what I have called the principle of hierarchic construction. Linguistic development in Brown's Stages II-V can likewise be described as a process of hierarchic construction. In Stage I the meanings expressed by *to Mommy, for Mommy, Mommy's,* are not yet fully differentiated. In Stage II this kind of differentiation-plus-integration takes place for the basic meanings of Stage I. In English such "modulations of meaning," as Brown aptly describes them, are accomplished through noun and verb inflections, prepositions, articles, and copulas. "All these," Brown writes, "like an intricate sort of ivy, begin to grow up between and upon the major construction blocks, the nouns and verbs, to which Stage I is largely limited." In Stage III the child completes the differentiation of sentence modalities (declarative, interrogatives of various kinds, negative, imperative) begun in Stage I. In Stage IV sentences are constructed in which other sentences function as units (*I hope I don't hurt it*). Finally, in Stage V two or more simple sentences are integrated into a single sentence, sometimes with the deletion of redundant components (*John bought some gum and a book*). All these kinds of development may be subsumed under Brown's "law of cumulative complexity," which I have translated into the language of hierarchic construction.

But the principle of hierarchic construction provides more than a convenient vocabulary. It also imposes rather severe constraints on theoretical interpretations of language and its development. To see how these constraints arise, let us consider Stage I language and some of the attempts that have been made to characterize it.

Piaget has long maintained that language builds on already existing structures fashioned during the period of sensorimotor intelligence that occupies the first 1½–2 years of life. In *A First Language* Roger Brown strongly supports this view:

> In sum, I think that the first sentences express the construction of reality which is the terminal achievement of sensori-motor intelligence. What has been acquired on the plane of motor intelligence (the permanence of form and substance of immediate objects) and the structure of immediate space and times does not need to be formed all over again on the plane of representation. Representation starts with just those meanings that are most available to it, propositions about action schemas involving agents and objects, assertions of nonexistence, recurrence, location, and so on. But representation carries intelligence beyond the sensori-motor. Representation is a new level of operation which quickly moves to meanings that go beyond immediate space and practical action. [Brown (1973:200)]

Brown emphasized that sensorimotor structures provide the basis for *meanings,* not grammatical categories or relations. Indeed Brown and other psycholinguists have had much more success in characterizing Stage I language from a semantic point of view than from a syntactic one. For example, Brown has shown that many of a child's first utterances express three "operations of reference," which he calls "nomination" (*see, there*), "recurrence" (*more*), and "nonexistence" (*all gone*). These operations are all present at the sensorimotor level. Again, Brown finds that the majority of utterances in Stage I express a small number (8–15) of two-term *semantic* relations: agent and action, action and object, entity and location, possessor and possession, etc. Finally, several authors have had some success in constructing generative grammars for Stage I language, using semantic categories and relations in place of syntactic ones.

Examining these semantically oriented studies from the standpoint of hierarchic construction, one is struck by the following anomaly. The meanings used to analyze Stage I speech—agent, action, instrumental, benefactive, and the like—are well beyond the Stage I child's level of abstraction. In fact they lie at a level of abstraction that most adults would probably find difficult to cope with. Unless one wishes to claim that such meanings are innate— and Brown rightly rejects this hypothesis as unnecessary—one must explain how these semantic categories can organize the utterances of a Stage I child. One might speculate that the Stage I child has certain vague and rudimentary notions corresponding to the adult categories. But what are these vague meanings and how do they arise?

By the end of the sensorimotor stage a child has constructed schemata that enable him to structure his experience in terms of objects and actions. His first words serve to integrate objects or actions into classes, or to label classes that he has already constructed. In either case the subjective meaning of a word is determined, I suggest, by the class of objects, actions, or objects-plus-actions that the child has associated with that word in the past. The child's partitioning of his (already highly structured) experience into classes labeled by words exemplifies the basic process of differentiation-plus-integration: the formation of a class is an action of integration based on perceived (but of course unarticulated) similarities, and at the same time an action of differentiation based on perceived differences. The similarities and differences underlying the construction of object-action classes are idiosyncratic and, for any given child, they are continually changing. Some of a child's classes (and hence meanings)—those that hold special interest for him—may be narrower than a typical adult's; most are much broader, because the similarities that unite large classes tend to be more salient than those that serve to differentiate these classes into subclasses. (The adult's classes have in fact arisen through successive differentiations.)

In short, the subjective meanings of object-action words evolve as the corresponding classes grow and become differentiated. This much seems fairly obvious. Less obvious, perhaps, is the assertion that *the meanings of relations develop in a precisely analogous manner.* Just as the meaning of an action-object word is defined by the class of its referents, so (I claim) the meaning of a two-term relation is defined by a class of *pairs of referents.* The meaning of the possessor-possession relation, expressed in English by the formula *A's B,* is defined for a given speaker by a certain class of pairs in specific contexts—in this case, because the relation is asymmetrical, *ordered* pairs (A, B). The phrase "possessor-possessed" merely designates the class; it does not define its meaning.

The preceding highly schematic account of meaning suggests the following developmental scheme for Stage I.

Having constructed an adequate collection of simple object-action classes, the Stage I child begins to construct *pair classes.* (Of course, the construction of simple classes continues.) The referent of a two-word utterance is an integrated construct. It arises through differentiation-plus-integration of simple object-action classes. Initially all such integrated pairs may make up a single pair-class. The meaning of this class, expressed in adult language, is something like "A and B are closely related aspects of the single object or action or object-action earlier designated by A and B or by both A and B." (Needless to say, such translations can never be very precise. Because they use a more differentiated vocabulary to convey less differentiated meanings, they always convey inappropriate distinctions.) In the course of its growth, the initially constructed pair-class undergoes successive differentiations. Ultimately these differentiations (and the corresponding differentiated meanings) are expressed by the syntactic and phonological devices peculiar to the linguistic environment (word order, noun and verb inflections, prepositions, stress, intonation, etc.). According to Brown's account, some of this development takes place in Stage I (for example, among children learning to speak English, the construction of *ordered* pairs), but most of it takes place (by definition) in Stage II.

By how much does the differentiation of pair-classes and the corresponding differentiation of meaning precede the child's use of the grammatical devices that mark these differentiations? Perhaps psycholinguists can devise a way to answer this question empirically. In the absence of empirical evidence, it is tempting to hypothesize that no *substantial* time-lag *necessarily* occurs between the internal processes and their overt linguistic manifestations. If this hypothesis is correct, one should be conservative in assigning meanings to two-word utterances that lack syntactic or phonological markers. In the absence of evidence for differentiation of meaning, the present theory offers little support for the common practice of using adult semantic categories to classify two-word utterances in Stage I. Still less does it support the construction of formal generative grammars for Stage I speech. All semantic classification schemes and generative grammars for Stage I speech have in fact failed to describe substantial portions of the very corpora on which they were based. This is the outcome one would anticipate for attempts to represent relatively undifferentiated meanings (defined by relatively idiosyncratic classes) by relatively differentiated meanings (defined by culturally stereotyped classes).

I suggest that adults tend to overinterpret children's speech. When a two-year-old says "Down!" does he mean "I want to get down" or "Please put me down"? Neither, because he has probably not yet begun to differentiate modalities, not to mention the other semantic categories that the adult versions of his utterance presuppose. Analogously, when a two-year-old says "Put floor" we infer from the context that he means "Put the book on the floor." If this, or something like it, is what he really does mean, then his omission of the word "book" calls for an explanation. I suggest, however, that the two-year-old may really mean "Put floor"—that that two-word utterance represents the meaning he has constructed.

The preceding discussion is meant to illustrate (rather than demonstrate) the hypothesis that linguistic development is a process of hierarchic construction based upon previously constructed schemata, in which meanings and the means of expressing them are developmentally intertwined. In an account of language acquisition based on the principle of hierarchic construction there is no need —and indeed no room—for innate syntactic, phonological, or semantic knowledge. What is innate is the physiological capacity for the specific processes of hierarchic construction involved in the growth of meaning and of linguistic competence. Children do not have an innate knowledge of principles of universal grammar (if such a thing exists). What they do know—what they have been programmed by biological evolution to know—is how to construct a natural language—any natural language. And that knowledge is infinitely more useful than a knowledge of constraints so abstract and general that no one has yet been able to identify them with precision. In constructing his first language a child is following a trail that has been cleared and blazed by those who have gone before him. There are many such trails, all radiating from a common starting-point and all in a state of continuous reconstruction. They are all different, yet they are all recognizably the same, because they are all products of the same process.

VII. *Conclusion: Intelligence, Culture, and Society*

In the preceding paragraphs I addressed two distinct problems: the evolution of interactive traits and the development of creative behaviors. I argued that, contrary to prevailing theoretical ideas, the fitness of an interactive trait depends in general on the distribution of homologous traits in the population, and that neither the emergence nor the stability of an interactive trait is governed, in general, by its inclusive fitness. As for behavior itself, I argued that, although learning plays a subordinate role to genetic programming in most kinds of animal behavior, the reverse is true for certain kinds of human behavior. Typical of, and of central importance among, such behaviors is the acquisition and use of language. I argued that what is innate in the human capacity for language is not a set of constraints—grammatical or semantic—on languages as such, but a capacity for constructing a language in a certain way. Any language constructed in this way is a natural language. A child acquiring his first language does not abstract or induce syntactic and phonological rules from linguistic experience, nor does he use linguistic experience to narrow down a set of genetically predetermined choices. Using clues provided by experience, he constructs a new language that more or less resembles the languages of family members and friends.

I shall now try to put these two arguments together.

If we compare human societies with other primate societies we are at once struck by the disparity in size. Among nonhuman primates, baboons have the largest social groups. A baboon troop typically contains about 60 members. Groups with 150 or more members seem to be unstable against fission into smaller groups. The smallest stable groups contain 15-20 members. These data, for which I am indebted to Irven de Vore, suggest (a) that the genetically programmed interactive behaviors of baboons confer their greatest benefits in communities substantially larger than the nuclear family, and (b) that these benefits do not increase indefinitely with group size. The second conclusion is emphatically not true for human societies. Modern history suggests that the benefits to individuals of increasing group size—for example, through the formation of multinational economic communities—increase indefinitely.

From a biological standpoint it is the human capacity to generate, store, and transmit information that makes possible our huge, intricately organized social groups. Thus civilization depends on the human capacity for creative behavior. But it is also true—and this perhaps is not so obvious—that creative behavior—behavior that improvises new solutions to unforeseen problems—depends on social organization and cultural transmission. For in an evolutionary context creative behavior is directly competitive with genetically programmed behavior. As we saw earlier, the two strategies are mutually incompatible. We lack the termite's innate knowledge of architecture; he lacks our capacity for improvising structures. But a behavior improvised on the spot will rarely be as effective as one that has been shaped by natural selection over thousands or tens of thousands of generations. It is the ability to store and transmit information that makes up for this deficiency and tips the balance in favor of improvisation. Cultural transmission allows us to combine the advantages of intricate, highly adapted behaviors built up over many generations with the capacity for innovation in the face of new environmental challenges and opportunities.

From these considerations my former student John D. Kelly has

252   DAVID LAYZER

concluded that intelligence (the capacity for creative behavior) and society must have evolved together, for they could not have evolved separately. Without the capacity for improvising behavior and transmitting information about improvised behaviors from generation to generation, social groups must remain relatively small and undifferentiated. But intelligence in the hypothetical state of nature envisioned by Hobbes, Locke, and Rousseau would be maladaptive. Lacking both the genetically programmed behaviors of less intelligent animals and the compensating benefits of culturally transmitted knowledge, an intelligent but nonsocial animal could not long have survived. In short, man is a social animal because he is intelligent and he is intelligent because he is social. Intelligence and sociality, the two most distinctive facets of human nature, are two sides of the same coin.

BIBLIOGRAPHY

Brown, R. 1973 *A First Language: The Early Stages.* Cambridge: Harvard University Press.

Chomsky, N. 1965 *Aspects of the Theory of Syntax.* Cambridge: MIT Press.

Dobzhansky, T. 1970 *Genetics of the Evolutionary Process.* New York: Columbia University Press.

Haldane, J. B. S. 1932 *The Causes of Evolution.* New York: Longmans Green.

Hamilton, W. D. 1964 *J. Theoret. Biol.,* 7:1.

——— 1975 "Innate Social Aptitudes of Man: An Approach from Evolutionary Genetics." *In,* R. Fox (ed.), *Biosocial Anthropology.* New York: John Wiley.

Layzer, D. 1978a *J. Theoret. Biol.,* 73:769–88.

——— 1978b *J. Social Biol. Struct.,* 1:297–305.

——— 1980 "Genetic Variation and Progressive Evolution." *Am. Nat.* (in press).

Mayr, E. 1963 *Population, Species and Evolution.* Cambridge: Harvard University Press.

——— 1974 *Amer. Scientist,* 62:650–59.

Montagu, A. 1976 *The Nature of Human Aggression.* New York: Oxford University Press.

Piaget, J.   1936   *The Origins of Intelligence in Children.* New York: Norton.

——— 1978   *Behavior and Evolution.* New York: Pantheon Books.

Schmalhausen, I. I.   1949   *Factors of Evolution: The Theory of Stabilizing Selection.* Philadelphia: Blakiston.

Slobodkin, L. B.   1964   *Amer. Scientist,* 52:342–57.

Stebbins, G. L.   1969   *The Basis of Progressive Evolution.* Chapel Hill: University of North Carolina Press.

Trivers, R.   1971   *Q. Rev. Biol.,* 46:35.

Waddington, C. H.   1961   *The Ethical Animal.* London: Allen and Unwin.

Wilson, E. O.   1978   *On Human Nature.* Cambridge: Harvard University Press.

S. L. WASHBURN

HUMAN BEHAVIOR
AND THE BEHAVIOR OF OTHER ANIMALS

Over the last few years there has been a great increase of interest in animal behavior. This started with popular books, articles, and television programs. Concern with the environment and ecology played a part. The airplane greatly increased opportunities for travel, both for the individual and the television camera. Popular books came first, well before the popularity of animal-behavior courses became evident on the campus. Robert Ardrey's (1961) *African Genesis* was particularly influential, but it was years later before textbooks on animal behavior were available in quantity. For example, the texts by John Alcock, Jerram Brown, Irenäus Eibl-Eibesfeldt (second edition), and Edward O. Wilson all appeared in 1975. I stress the popular roots because the apparently sudden interest in animal behavior cannot be understood if the academic scene is considered in isolation. Before World War II there was almost no interest in the naturalistic behavior of the nonhuman primates, and the late C. R. Carpenter was deeply disappointed by the reception that his monographs on the behavior of the howler monkey (Carpenter, 1934) and the gibbon (Carpenter, 1940) received.

Many of the popular books—particularly those by Robert Ardrey (1961), Desmond Morris (1967), and Konrad Lorenz (*On Aggression,* 1963)—stressed the importance of animal behavior and

From *American Psychologist,* Vol. 33, No. 5 (May 1978). Copyright © 1978 by the American Psychological Association, Inc. Reprinted by permission.

evolution in the understanding of human behavior. This emphasis is carried to an extreme in sociobiology and may account both for the great interest in and the resistance to this science. As I see the situation, animal behavior is both fun and science. It is fascinating to learn of the language of the bees, of pheromones, of factors guiding migration in birds and fish, and of the life of the penguin or the world of the herring gull. The emphasis on experiments in recent studies of behavior led to unexpected advances, but the studies were fragmented, and there was little emphasis on theory, social behavior, or evolution. Wilson's *Sociobiology* provided a new emphasis, synthesis, and theoretical foundation; it added the excitement of the promise of a new science. If *Ethology* (Eibl-Eibesfeldt, 1975) and *Sociobiology* (Wilson, 1975), both of which appeared in the same year, are contrasted, the new emphasis may be clearly seen.

It might be helpful if a science historian would record these changes in the study of animal behavior while they are still in progress and while the relevant people are alive—the various factors might now be sorted out in a way that will not be possible many years hence. My opinion is that the speed of acceptance of sociobiology shows the need for some such theory, shows that the need is evident in our whole climate of opinion—both popular and scientific—and shows that the controversies of the last two years cannot be understood in terms of the progress of science alone.

Animal behavior is worthy of study in its own right, even if it were never applied to human beings, but the present controversies stem from the use of animal behavior to understand human behavior. The descriptions of the sociobiological controversies in *Bio-Science* (Sociobiology Study Group of Science for the People, 1976) and *Science* (Wade, 1976) show this to be the case. The accounts in *Time* ("Why You Do What You Do," 1977) and *Psychology Today* (S. Morris, 1977) indicate a strong popular interest in biological interpretations of human behavior, a thesis with a long history in western European culture.

In the briefest terms, sociobiology is a theory that seeks "to develop general laws of the evolution and biology of social behavior" and to extend such laws to the study of human beings (Wilson 1977b: xiv). It builds on natural selection in its modern form (the synthetic theory, shaped by about 1940), with emphasis on behavior (fully stated by Roe & Simpson, 1958), and adds the theory of kin selection (Hamilton, 1975). Natural selection states that the biology that leaves the most offspring is favored; kin selection shows that altruistic behavior may also be selected for, if it benefits individuals closely related to the altruist. This added element makes possible a much closer fit between evolutionary theory and social behavior—and it is this application to a wide variety of human behaviors that has caused the reaction.

The issues are complex, arising in part from the extreme claims of sociobiology and in part from the long history of biological science that has confused and retarded the development of the social sciences. While sociobiology is new, especially in its enthusiasm and emphasis on social behavior, it also continues a history of scientific error. When applied to human behavior, it renews the mistakes of social Darwinism, early evolutionism, eugenics, and racial interpretations of history. To defend its position sociobiology must (a) attack social science, (b) minimize the difference between learning and inheritance, and (c) postulate genes for a very odd assortment of behaviors—altruism, cheating, spite, deception, creativity, and reciprocity, to mention only a few.

Postulating genes for behaviors, the most controversial part of the application of the theory, appears to repeat the mistakes of the eugenicists, who postulated genes for alcoholism, crime, and other behaviors of which they did not approve. Medawar and Medawar (1977: 38) have described this kind of thinking as "geneticism," which they define as "the enthusiastic misapplication of not fully understood genetic principles in situations to which they do not apply."

So sociobiology may be described as new, challenging, and destined to dominate the future thinking of both biological and social science, or it may be described as mostly old and given to repeating errors of the past. As shown by statements in the press and in reviews in magazines and journals, emotions are deeply involved in this controversy. When this is the case, I think of Paul MacLean's (1970) statement:

> Emotional cerebration appears to have the paradoxical capacity to find equal support for opposite sides of any question. It is par-

ticularly curious that in scientific discourse, as in politics, the emotions seem capable of standing on any platform. Different groups of reputable scientists, for example, often find themselves in altercation because of diametrically opposed views of what is true. Although seldom commented on, it is equally bewildering that the world order of science is able to live comfortably for years, and sometimes centuries, with beliefs that a new generation discovers to be false. [p. 337]

In this situation, particularly because of the deep popular interest, there will be no simple solution to the conflicts. What I hope to do is present a series of examples that may clarify some issues. In the interest of brevity, the examples have to be oversimplified, and I hope you will bear with me as I strive for both accuracy and condensation.

Behavior of the Apes

In the latter part of the 19th century, the behavior of the great apes was the starting point in many evolutionary sequences. There were scientific controversies over promiscuity, monogamy, and many other issues. A very large number of papers had been written, and Yerkes and Yerkes (1929) compiled a major summary (over 500 references) in *The Great Apes*. Yerkes and Yerkes concluded (p. 590): "As one reflects on the situation it appears first incredible, then ludicrous, that as professional scientists we should depend on accident instead of intelligently planned and prearranged conditions for the extension of knowledge and the solution of significant problems." But is it not even more ludicrous, in retrospect, that the nature of ape behavior and its relation to human behavior was debated for more than 100 years before there were any reliable data? Robert Yerkes was a major figure in promoting field studies and, though a careful investigator, concluded that the social unit of chimpanzees "is undoubtedly the family" (p. 541), and that gibbons lived in herds of up to 50 animals. C. R. Carpenter and subsequent field-workers have shown that the social unit of gibbons is a male-female pair. Even Yerkes, when he described the situation as "ludicrous," could not believe that the data were as ridiculous as they really were. A comparison of the social behavior of Bigfoot and the Abominable Snowman would be as useful for the study of human evolution!

There are at least three reasons why scientists accepted accounts of the behaviors of the great apes long before there were any reliable data: (1) Animal behavior was considered to be simple and instinctive—therefore anecdotes, stories by local people, and travelers' tales were considered useful. Hunters particularly were credited with knowing a great deal about animals. (2) The theory of evolution so dominated science and popular opinion that the misinformation had to be put in an evolutionary context. The behaviors of highly diversified animals were put in a few simplified stages. (3) There was so much faith in the power of the theory of evolution that it was believed that the theory could compensate for research and careful verification of information. But a theory is only a license for research. Natural selection, a great advance over theories of special creation or evolution by acquired characters, is an advance only because it guides research into more productive channels. There is no way in which the theory of natural selection can substitute for careful research on the behaviors of the animals under consideration.

The beginning of reliable studies on the natural behavior of the great apes may be marked by George Schaller's (1963) study of the gorilla and Jane Goodall's (1968) long-continued investigations of the chimpanzee. The results of these studies are facts of an entirely different order from the anecdotes of the last century. The traditional gorilla was a fearsome beast, but, of course, he was a wounded gorilla facing a brave hunter—a vision the 19th century wanted. It was replaced by the perception of the peaceful gorilla and the friendly chimpanzee, views stemming from the conditions of the early fieldwork and the human desires of the 1960s. Today chimpanzees are seen as aggressive, territorial, dominance-seeking animals; they have been observed in an act involving the killing and eating of an infant chimpanzee from another group (Goodall, 1977). Gorillas fight and on occasion may also kill other gorillas —they too have been observed eating a young gorilla. Audiences who heard talks by Dian Fossey in 1977 were upset by her description of gorilla violence. In her talk at the Leakey Foundation, Galkidas-Brindamour (1977) noted that all old male orangutans have scars and signs of past battles, presumably from struggles with other orangutans. This view of the apes is contrary to the present climate of opinion, and it will be interesting to see how the conflict between behavioral fact and popular desire is mediated.

Goodall and Hamburg (1975) have summarized the evidence for using the chimpanzee as a model for the behavior of early man. The quality and quantity of the information is impressive compared to that in earlier accounts, but the facts still depend on the fieldwork of a very few people in a few localities. It is still uncertain how much additional fieldwork will be necessary before chimpanzee social behavior and ecology are adequately understood. Even less is known about the behaviors of the gorilla and orangutan, so although the data are vastly better than was the case when Yerkes and Yerks wrote *The Great Apes,* human understanding of these primates is still very much in process. Even with the aid of comparative anatomy, the fossil record, and data from the study of captive animals, major problems remain, and it will be some years before intellectually satisfying comparisons between human behaviors and those of our nearest relatives will be possible.

An example may make the issues clearer. Sociobiologists have revived the old theory that the loss of estrous behaviors in human females is a major factor in giving rise to the human family. The human family has been described as a "pair-bond" whose consistency may be aided by the loss of estrus (Barash, 1977b: 293, 295, 297). Ronald Nadler (1977), however, describes the shortest estrus and the greatest limitation of mating behaviors in the gorilla —the most continuously social of the great apes. Orangutans will mate at any time because the males mate whether the females are in estrus or not; yet the orangutans are the most solitary of the great apes. Gibbons form lasting pair-bonds but are the least sexually active of the apes. Further, many monkeys have restricted mating seasons, but the social group continues throughout the year, as in rhesus monkeys (Lindburg, 1977). Maximum fighting among males occurs during the mating season, and many monkeys may be killed (Wilson & Boelkins, 1970).

The idea that sexual attraction is responsible for the existence of the social group in nonhuman primates and for the family in man is appealing. It is simple and easily stated, and in our culture, biological explanations are regarded as more scientific than social ones; but the sexual attraction theory of society does not fit the data—even in the case of our closest relatives. Whether monthly or yearly cycles are considered, social systems continue to function. The importance of sex as *the* binding mechanism is an idea that comes from 19th-century European culture, just as the lack-of-estrus–monogamy theory comes from a simplistic view of our culture. Social organization is the most important adaptive mechanism, and sex adds a wide variety of attractive or disruptive behaviors. Adequate data are only now becoming available in 1977, although the subject has been discussed for over 100 years.

As far as the field data go, male–female relations in nonhuman primates may vary from near-promiscuity, to consort relations of varying duration, to lifelong pair-bonds. The interpretation of such relations is complex, and the behavioral classification is only a first step. In the case of gibbons, which do form lasting male–female pairs, the interpretation of the pattern, at a minimum, involves factors of locomotion, territory, diet, teeth, calls, aggression, and the brain (Washburn, Hamburg, & Bishop, 1974). In even this simplest case of a primate social system, adaptation is complex, and a great deal of additional research has to be done before anyone can have much confidence in the interpretation.

Sociobiologists reduce human social behaviors—the extraordinary variety of human male–female customs—to monogamy, or, even worse, pair-bonding, but even the word *monogamy* does not stand for any clearly defined behaviors that could have a genetic base. Monogamy may mean a continuing relationship, a temporary one, or it may be modified by all sorts of extensions and exceptions that have biological consequences (concubines, slaves, mistresses), and these are combined with the widest sort of variation in economic, social, and sexual practices.

The term *polygamy* is open to similar but even greater objections. It may refer to the marriage of one man to more than one woman, or to the marriage of one woman to more than one man. The number of polygamous arrangements in human societies is exceedingly large and must have a variety of genetic consequences. The important point about human monogamy or polygamy is that these terms refer to systems of marriage, not mating. Mating can and does occur outside of marriage in every known society; marriage is a socially recognized arrangement involving rights and duties and encompasses far more than sexual behavior. This fundamental difference between the traditional and reciprocal customs of humans and the mating behaviors of other animals appears to be totally ignored by animal behaviorists. The use of such words by sociobiologists shows a total misunderstanding of social science. Even ape behavior is far too complicated to be analyzed by labels and guessing.

Sociobiologists speak of parental investment as if this were something that could be predicted on the basis of genetics, but, to take a single illustration, note how different parental investment was in the pastoral Masai culture and in the France of a century ago, where maintaining the small farm was critical. Customs are imbedded in complex social patterns, and it is the patterns that must be compared if the customs are to be understood.

Molecular Taxonomy

One area in which there has been great progress is in the classification of the primates. Recent progress in the study of biochemical evolution (DNA, sequence data, and immunology—reviewed by numerous authors in Goodman & Tashian, 1976) shows that humans, chimpanzees, and gorillas form a very closely related group. Orangutans are considerably less closely related, and Old World monkeys are far less similar. These relationships, futilely debated for many years, are in my opinion now settled, and I believe that a quantitative molecular taxonomy will be a major factor in providing a framework for comparative studies. King and Wilson (1975: 108) conclude that "the sequences of human and chimpanzee polypeptides examined to date are, on the average, more than 99% identical."

Polypeptides evolve at different rates, but the rates are relatively constant for each kind of polypeptide. Therefore, it is possible to estimate, from the molecular data, the times at which various lineages separated (Wilson, Carlson, & White, 1977). The procedures are still controversial, but I think it will be possible to estimate

262   S. L. WASHBURN

divergence times more accurately from the biochemistry than from the fossil record—particularly in the case of groups, such as the primates, for which the fossil record is fragmentary.

Speech

Studies about the communication of many mammals, birds, and insects have revived interest in the old and important problem of the origin of talking. The studies of ape communication have been particularly exciting and have opened a new area for scientific investigation. All of these studies raise fundamental issues on the meaning of words and the nature of comparison and evolution.

Human speech is based on a phonetic code, a system in which short, meaningless sounds (phonemes) are combined into meaningful units (morphemes, words). This system has the characteristic of codes in general—a relatively small number of units may be combined in an almost infinite number of ways. So the human-sound code system of communication is open, and even with the needs of modern science, no limit has been approached.

The behavioral problem has three aspects: (1) to understand the human system, (2) to compare it, and (3) to speculate on its evolution. If we proceed in this manner, the first discovery is that the phonetic code is unique to our species. There is no comparable behavior in any other primate, or any other animal for that matter. Apes cannot be taught to talk in spite of great efforts (Kellogg, 1968). R. Allen and Beatrice Gardner and those who have followed their lead have wisely shifted from communication by sounds to a different anatomical system. Speech, however, is the primary form of human communication, and if the aim is to understand speech, the phonetic code is the issue—the power to generate new words to meet any need stems from the code.

It should be remembered that the phonetic code is not obvious. Words appear to be the important and meaningful units, and when people started to write, the first form of writing used signs for words. This requires an enormous number of signs, whether in writing or in gesturing. It took some 2000 years for the people in the Near East to find that a few letters symbolizing sounds could do the job of thousands of hieroglyphs. Millions of man-years have been wasted because the phonetic code was not obvious. The Chinese had to learn more than 10,000 characters instead of two dozen letters. And the spelling of English is a reminder that many still do not understand the phonetic code—every schoolchild wastes countless hours memorizing spellings that should have been discarded long ago.

Some years ago, David Hamburg (1963) made the point that behaviors that have been important in the evolution of a species are easy for the members of that species to learn. That is, selection of the successful behavior led to selection for its biological base, so that, over time, the behavior became both more effective and easier to learn. Languages are easy for humans to learn—but such learning is impossible for other primates. Clearly, evolution has produced a biological base that makes this learning so easy for humans. In both New and Old World monkeys, the characteristic sounds of the species may be elicited by stimulation of the limbic system near the midline (Ploog, 1970; Robinson, 1967, 1976). Extensive removal of cortex does not affect sounds and affects facial expression only slightly (Myers, 1968). The anatomical evidence has been reviewed by Myers (1976), and it has been shown that in humans, the primitive system is still present, but the cortex has become very important in both speech and facial expression and provides the biological basis for learning in both.

In summary, the view briefly presented here is (a) that the phonetic code is the basis for the success of human communication, although supplemented by gestures, postures, and facial expression, (b) that easily learning the code is made possible by the brain, and (c) that some evolution in the articulatory mechanisms is involved. If this brief outline is accepted (at least for the moment), it may be used to illustrate some of the problems of comparing the behavior of human beings and nonhuman primates. There are several points:

1. Comparisons must start with human behavior, or reductionism is almost certain. The only way we know that a phonetic code exists is through studying human beings. If the study begins with monkey communication, for example, it can be observed that monkeys make some two or three dozen sounds (Struhsaker, 1967,

1975), and it can be speculated that more sounds will lead to speech. The monkey system of communication, however, uses both gestures and sounds in combination (Lancaster, 1975). The gestures are almost always more important, and when the sounds are separated and treated as an independent system, the monkey communication system is, in fact, destroyed. It is a very human bias that sounds should be so important or should constitute a separable mode of communication.

2. There has been an enormous amount of description and speculation, but very few experiments form the base for contrasting the limbic control in monkeys with the cortical control in our species. The critical experiments have not been performed with apes, but judging from learning, the structure of their natural communications will be like that of monkeys. The issue is that there are really two kinds of comparisons. First, sounds may be compared in all sorts of creatures. This has been done by Thorpe (1972) using 16 design features. This is fun—it calls attention to a wide variety of communicative behaviors—but if the purpose is thorough understanding of a particular comparison, it is essential to analyze the biological base experimentally. The weakness in the outline presented above lies in the experiments; there have not been enough comprehensive studies done on enough primates.

3. In the analysis of almost any particular problem, the evolutionary order should be considered last. More is known about human beings than about other animals; if comparisons start with our species, the problems are more likely to be usefully stated. It is almost inconceivable that, starting with the extraordinary diversity of human behavior, the human family would be called a "pairbond." The fossil record and archeology are essential to give a perspective on time and the diversity of previous forms. But the record is very incomplete, and the study of evolution involves uncertain reconstruction and the evaluation of many different opinions.

4. There is a biological basis for speech. It is the human brain that learns language, but there is no evidence that the brain determines the particular language learned. Any normal human being learns the language of its group, and a second if necessary. This may be the clearest example of the nature of many biosocial problems. The biology is basic—it sets limits, gives probabilities, and

quite possibly accounts for some linguistic universals. But the freedom of learning is immense, and constraints are primarily in the learned systems, not in the biology.

In Africa there were hundreds of mutually unintelligible human languages in an area in which baboons used the same system of communication. There were over 100 tribal groups in what is now Tanzania (Bienen, 1970), an area in which there is no evidence for any major behavioral differences in baboons.

The ease with which human languages are learned and the speed with which they change distinguish the behavior of *Homo sapiens* in a way that is characteristic of no other species. In the nonhuman primates there may be minor regional differences in communication, just as there are dialects in birds, but nothing comparable to the diversity of human languages. Language is probably the best example of a common biological base facilitating the learning of an extraordinary variety of behaviors.

Fashions have moved back and forth on the nature–nurture problem. Traditional social scientists stress learning. The sociobiologists attack the distinction, pointing out that both are always present. From a practical point of view, this is not one problem; some behaviors may be largely genetically determined, and for others the environment may be important. Granted basic human biology, there is no evidence for any genetic determination of any particular language.

Language is the behavior that distinguishes human social behavior from the social behavior of the other primates. As the child learns the language of its group, it learns how this system of communication is used in social relations and in economic, political, and religious systems. The social systems of nonhuman primates are based on the biology of the species, observational learning, and interaction. To these the human species adds the social complexity made possible by brain, language, and technical advance. Viewed in this way, biology supports two of the most fundamental assumptions of the social sciences—that human populations, not individuals, have the same behavioral potentials, and that social systems are not determined by the basic human biology. This, then, is the reason sociobiologists and others seeking genetic explanations of human social behavior *must* minimize the difference between learn-

266 S. L. WASHBURN

ing and inheritance (nurture and nature) and attack the notion that social facts should be explained by other social facts (Wilson, 1977a).

It is language (cognition, speech, articulatory mechanisms) that gives the nature–nurture, genetic–learning controversy a wholly new dimension in human behavior. There is no comparable mechanism, nor any alternative biological system, that allows almost unlimited communication and the development of new symbols when needed.

Social science might be defined as the science that studies the nature, complexity, and effectiveness of linguistically mediated behaviors. Such a definition clearly shows why human social behaviors cannot be reduced to those of other animals, even those of our closest living relatives. Or, alternatively, if one puts the matter positively, it is the evolution of new structures in the human brain that forms the basis for learning language (speech and cognition); so the uniqueness of human behavior may be defined behaviorally or by the biology that makes the behaviors possible. The significance of the biology can only be seen by comparing the behavior of the animals to that of humans. In the primates, the difference is reflected in communication systems that range from one that may be described in a very few pages to one that is only partially described in dictionaries thousands of pages long.

As Alexander (1974:376) has stated, "human social groups represent an almost ideal model for potent selection at the group level." The reason for this is that human groups adapt through knowledge and organization, both of which depend on speech. Knowledge and organization are properties of groups and are transmitted from generation to generation by learning, not genes.

Knowledge and organization, dependent on speech and intelligence, make possible technical progress. The differences between the simplest sort of hut and a modern high-rise building are the result of purposeful human effort, not blind chance. Possibly the greatest difference between biological evolution (changes in gene frequencies over time) and history (human learned behavior over time) is that biological evolution depends on mutations (producing variation that is edited by natural selection), while technical progress is the result of intelligent purpose.

Walking

In discussing the biological basis for speaking, the analysis is limited to comparisons of living creatures. In spite of many ingenious attempts, no way has been found to prove what kind of communication characterized our earliest ancestors. In the case of walking, there is a substantial fossil record, and bipedalism was present more than 3 million years ago. At that time, our ancestors had brains no larger than those of the contemporary apes (Washburn, 1959). The teeth were very human in form, and these early men (genus *Australopithecus*) had probably been using tools extensively for some millions of years, although the earliest stone tools do not appear in the record until 2.5 million years ago (Isaac, Harris, & Crader, 1976). It appears that the evolution of a new pattern of locomotion was the behavioral event that led to the separation of human beings and apes, and this may well have been in a feedback relation with object-using and hunting. The fossil record allows the determination of an order that could not be deduced from the study of the contemporary forms. It is likely that the conclusion of greatest psychological interest is that the large human brain evolved late—millions of years after the separation of man and ape. The human brain is unique because it evolved in response to the selection pressures of the latest phases of human evolution.

However, the fossils alone would give very little indication of the magnitude of the changes correlated with the new locomotor patterns. Not only did the bones change, but there was major reorganization in the muscles, far more than might be inferred from the bones. So interpreting the differences in the locomotion of human beings and apes involves three kinds of comparisons.

First, the behaviors of the contemporary animals show what they actually do. Second, study of the fossils puts a time scale and order into the comparison. Third, analysis of the structure and function of the contemporary forms shows how complex the changes were. Since human locomotion is far better understood than that of any other animal, and because it is in many ways unique, it is useful to start the comparisons with what is now known about human beings, then to shift to comparisons with other animals, and finally to resort to the essential but always incomplete fossil record. The observation of behavior (for example, knuckle walking in the chimpanzee and human bipedal locomotion) gives no information on the very extensive internal anatomical reorganization that makes the behavioral differences possible. To the extent that locomotion is an essential part of human behavior, there cannot be an independent study of behavior, a study independent of major efforts to understand the internal mechanisms that are the basis for the externally viewed behaviors.

Through the process of natural selection, evolution produced a complex structural–anatomical base that makes human bipedal locomotion possible; yet the behavior has to be learned and it takes a child years to walk and run efficiently. Much has been made of the fact that it takes years to learn to speak—but motor learning takes equally long. In both cases the child wants to learn, and, under normal circumstances, learning cannot even be prevented. The basis for walking is common to the whole human species, and although there are minor anatomical variations, all human populations share the structures that make learning to walk, run, etc., possible. It is clear that there is a major, genetically determined biology that separates humans from all other kinds of animals, although nothing is known of the genes involved.

This situation for walking (structural base, propensity to learn, important unique behavior) may be contrasted with that for swimming. Apes cannot swim (Wind, 1976), and most humans never learn to swim. There is a biological base for swimming, but the structure of the arms (primarily the result of the arboreal life) and of the legs (evolved for walking on land) makes human swimming a peculiar, new activity that is not the result of evolution. Granted human intelligence, some humans may learn to use arms and legs in a variety of ways that may lead to swimming. There has not been selection for swimming; no genes different from those involved in other human adaptations are required. The behavior simply uses structures that evolved as parts of different functional–behavioral complexes.

This example shows that even when behaviors are based on well-understood biology, it is not possible to infer that special genes are correlated with the behaviors. Obviously, human hands may learn a very wide variety of behaviors, although the basic biology of bones, joints, and muscles is the same. Further, it should be noted that human beings elaborate on everything. People do not just walk and run; they hop, skip, dance, jump, and race. The same basic biology makes all these behaviors possible. Likewise, people may swim in many different ways, and there is no evidence that the elaborations are based on biological differences. The social sciences assume basic human biology and are concerned with the extraordinary variety of behaviors that may be learned. Human biology is primarily concerned with understanding the common biological base. The biological differences between individuals may be important in medicine and psychiatry or in achievement, but individual biological difference is rarely important to the social sciences.

Evolution

Evolution is a master theory. For a long time it dominated both social and biological science and it became a part of the general climate of opinion. It was influential in rationalizing the 19th-century world and proving that its customs were justified by a Darwinism that now seems more like a religion than a theory guiding research. Mayr (1972) has shown that it was difficult for people to accept the theory of evolution because it directly affected custom and belief. Once accepted, however, the theory dominated a very large part of human thought. As noted earlier, the original theory of natural selection was strengthened by the discovery of the genetic mechanisms, clarified as formulated in the synthetic theory, and further strengthened by the addition of kin selection. The theory has been clearly stated recently by numerous authors— Alexander (1974), Barash (1977a, 1977b), Hamilton (1975), and Wilson (1975), to mention only four who have particularly related the theory to the evolution of behavior and sociobiology. Present controversies stem not from the theory but from its applications and, particularly, from its applications to human behavior.

In simplified form, the issue is that the sociobiologists are so confident of the power of evolutionary theory that they think their conclusions must be accepted, or at least accepted with minor modification.

Hamilton (1975:150), a leader in the development of the genetic theory of kin selection, concludes that because benefits to fitness will not go to relatives in large societies, that "civilization probably slowly reduces its altruism of all kinds, including the kinds needed for cultural creativity." The absurd conclusion is a correct deduction from the evolutionary theory. It simply proves that the theory is useless when applied to the interpretation of learned behaviors and recent human history. There is no evidence that civilization reduces altruism—quite the contrary—and there is every evidence that cultural creativity is greatly increased. In his 1975 paper, Hamilton raises the fundamental issues by combining the latest genetic evolutionary theory with an application to human evolution that could have been written 100 years ago. No matter how powerful the theory of evolution, there is no way to go directly from the theory to the interpretation of human behaviors. The theory guides research. It does not provide conclusions.

The problems caused by the overconfidence in evolutionary theory may be illustrated by an example from archeology. The stone tools made by the Neanderthals were ordered by archeologists into several evolutionary stages, but when the assemblages of stone tools were dated by independent methods, some of the supposed stages proved to be contemporary. The "stages" may represent different cultural adaptation or, perhaps, only fashions. The reality of earlier traditions (Acheulian) was even more distorted by being arranged in orders from simple to complex without regard to the actual associations and order in the ground. Interest in evolution stimulated interest in archeology, but the science would have progressed far more rapidly if scholars had never heard of the theory of evolution and had been forced to determine the actual sequences of the stone tools rather than placing them in arbitrary evolutionary sequences.

Just as no facts on ape behavior were needed to make the evolutionary reconstructions, so no archeological facts were needed to understand the history of cultures. Even more dramatically and disastrously, the theory of evolution was thought to be so compelling that it gave people confidence to arrange anecdotes, travelers' tales, and trivial information into social sequences that never existed. The evolutionary orders were, in part, the creation of scholars, but they were accepted because they fitted the social and political climate of the 19th century. In the order Savagery, Barbarism, Civilization, one can guess where monotheism, monogamy, and European customs will be placed! As noted earlier, the acceptance of conclusions may be a far better guide to intellectual and social reality than the supposed proofs. Theories are necessary guides to research, but an accepted theory (such as the theory of evolution based on natural selection) may be a terrible liability if misapplied. All the prestige of the theory then supports the mistakes.

For example, Wilson (1975:551) gives rules for the reconstruction of phylogenies of behavior from the behavior of contemporary animals. The rules were derived from comparative anatomy, developed before there were many fossils, and they are the very part of comparative anatomy that does not work—or at least works only to a very limited extent. These are the rules that were used to create the missing links, but none of the reconstructed links looked like the fossils that were actually found. When people, both laymen and scientists, wanted missing links, they made them. As regards human behavioral evolution, there are fossils and a rich archeological record, but no way that the actual historical past can be deduced from the present. At one time it was believed that the brain evolved early and led in human evolution, but the record shows that bipedal locomotion evolved some millions of years before large human brains. While the evidence is less clear, it is probable that tool-using and hunting also preceded large brains by many hundreds of thousands of years. It is more useful to regard our brains as products of human evolution rather than as causes. Obviously, the brain and behavior were in a feedback relation.

To suggest that the evolution of human behavior should be reconstructed from the behavior of contemporary peoples is particularly unfortunate for two reasons. First, this is precisely what was done in the 19th century, and this is what created the mistaken sequences. We have lived through those mistakes. Second, in addition to written history, there is now a great deal of archeology that carries the direct evidences for human behavior back many thousands of years.

When people anatomically like ourselves (*Homo sapiens*) appear in the fossil record, history accelerated to a remarkable degree. Geographically, people crossed water and arrived in Australia. They conquered the Arctic and arrived in the New World. The development of boats, harpoons, fishing, bow and arrow, and many other behaviors are reflected in the record. Lenski and Lenski (1974, especially on p. 131) give a very useful review of the technical advances. The end of this era is marked by the domestication of the dog, one of the many indications of the cultural similarity of all the late populations of *Homo sapiens*.

In contrast to the earlier changes, measured in times of hundreds of thousands of years, these late changes do not seem to correlate with biological evolution, or at least not in any simple way. It appears from the record that, granted the basic biology of our species, learning was the dominant factor in determining behavior, even 30,000–40,000 years ago. Putting the matter somewhat differently, human cognitive abilities, speech, and social systems— that is, human nature as we know it today—had fully evolved before 30,000 years ago.

If this point of view is accepted, then the background to Cohen's (1977) major reappraisal of the origins of agriculture is the basic biology of *Homo sapiens* and the cultural consequences of this kind of biological organization.

With agriculture, a series of changes follow that affect the application of the theory of natural selection to human behavior. The more people there are, the more mobility, the more knowledge, and the more technology; and the shorter the time span, the less will genetic factors be important in determining human behaviors. With agriculture, human populations expanded *at least* 200 times, and history is full of documentation of trade with distant places, migrations, and wars. Sociobiologists clearly recognize that their view of the evolution of society depends on acts benefiting relatives (Hamilton, 1975) and that this becomes increasingly improbable as the conditions of the world changed after the agricultural revolution.

Ethology has been so important in the development of studies of animal behavior that a brief mention is essential, even though its distinctive characteristics are disappearing and it is merging into a much more general study of animal behavior (Bateson & Hinde,

1976). The genius of ethology was not the study of natural behavior; it was the design of imaginative experiments analyzing the behavior. Ethologists devised methods for determining what the animal was responding to, and this led to a science far from the early anthropomorphic studies of animal behavior. Unfortunately, these behavioral experiments led to postulating physiological mechanisms rather than to further experimentation. Interestingly, Barash (1977b:6) severely criticizes their methods, but later (p. 277) he praises human ethologists for using precisely the same methods! Surely the concepts that are inadequate for animals with relatively simple nervous systems will be even worse when applied to speaking, culture-bearing human beings.

Human ethology, a growing and enthusiastically supported branch of the field, is based on the direct observation of behavior, free of experimental intervention. *Human ethology might be defined as the science that pretends humans cannot speak.* Even infants respond to sounds long before they can talk, and mothers live in cultures where the customs, including how to take care of children, are transmitted in large part by speaking. A long history shows that the problems of human behavior that can be effectively studied by observation alone are very limited. Clearly, this is an extreme case of the study of human behavior being limited by the drawing up of the rules on the basis of the behavior of other animals. As noted before, a rich study of *human behavior must start off with human beings;* otherwise, critical behaviors are lost. Human ethology is an extreme example of a science not adjusting to uniquely human problems.

The human cortex is far more important in controlling human facial expression than is the case in monkeys. A minimum of experimental and clinical information suggests that human expressions are not reducible to a small number of biologically controlled, universal kinds. The ethological emphasis on classification rather than on internal biological structure precludes the possibility of usefully comparing human facial expressions with those of the nonhuman primates.

Ethologists often compare the behavior of highly diversified animals to that of human beings, often to the behavior of a single tribe. This mode of comparison (aggression in insects, fish, birds, lions, wolves, apes) is interesting. It raises questions. I like Michael Chance's expression that the comparison *alerts* us to possibilities. But it is not proof.

Lorenz (1963) concluded, after reviewing aggressive behaviors in a number of animals, that "the main function of sport today lies in the cathartic discharge of aggressive urge" (p. 280). The conclusion is based on mistaken physiology, and Lorenz has been criticized for his conclusions in *On Aggression*. The point is that the problem is *the method,* not the particular book. Animal behavior is extraordinarily diversified, and almost any thesis may be defended if an author is free to select the behavior of any animal that supports the contention of the moment.

Students of animal behavior feel free to use the behaviors of nonhuman species when making points about human behavior. For example, in a recent book, the chapter on human behavior cites the behaviors of many nonprimates to make important points. The possibility of atavistic behaviors in human beings is illustrated by a picture of a musk-ox in a defensive position. To show how peculiar this habit of proof really is, consider what the reaction would be if I sent to a zoological journal a paper on the musk-ox with defensive positions illustrated by the British squares at the Battle of Waterloo! The editor would be surprised, to put it mildly, and yet this is an accepted mode of thought when animal behavior is compared to that of man.

Ethology has a very strong bias toward equating behavior with easily visible external behaviors (Blurton-Jones, 1975). Careful observation is essential, and visible behaviors may be organized in a scientific manner. As far as human behavior is concerned, Hinde's (1974) *Biological Bases of Human Social Behavior* is the most useful statement reflecting this point of view. But most of the examples of behavior are not of human behavior, but of birds and monkeys. The brain is not mentioned, speech is minimized, and the book contains little biology. A contrasting view of the nature of behavior is given in Altman's (1966) *Organic Foundations of Animal Behavior,* in which the emphasis is on understanding the internal mechanisms of behavior. The external view and the internal view supplement each other up to a point—but I believe that progress in understanding the human behavior and its relations to the behaviors of other animals will come through analysis and understanding of physiological and anatomical factors, primarily the brain. From this point of view, fieldwork and observation of behavior are necessary in setting the problems, but such science will remain superficial unless supported by the much more complicated analysis of the internal mechanisms. It should be stressed that the external and internal points of view are, in theory, supplementary. Unfortunately, in practice, they frequently are not.

Genetics

The theory of natural selection is a genetic theory. As such, it will be useful in guiding genetic research. If long time spans are involved, it will be helpful in guiding evolutionary research. As noted earlier, it is a great improvement over such theories as creation or the one that hypothesizes that new characters may be acquired through use or disuse. If the issues are behavioral and anatomical changes over very long periods of time, natural selection must be an important guiding theory. The present controversies stem from differing opinions on the importance of the theory for interpreting short-term changes.

The fundamental calculus in sociobiology is based on the genetic resemblance between relatives. It is included in almost every statement on sociobiology but is particularly clearly stated by Barash (1977b:85–89). Parents share one half of their genes with an offspring, full siblings share one half, and there is less sharing with more distant relatives. It should be noted that the genetic sharing decreases very rapidly—far more rapidly than social obligations in any human social system. Sahlins (1976) has shown that actual kinship behaviors in human societies do not follow the genetic calculus, but his book, so far, seems to have had little influence.

This whole calculus upon which sociobiology is based is grossly misleading. A parent does not share one half of the genes with its offspring; the offspring shares one half of the genes in which the parents differ. If the parents are homozygous for a gene, obviously all offspring will inherit that gene. The issue then becomes, How many shared genes are there within a species such as *Homo sapiens?* King and Wilson (1975) estimate that man and chimpanzee share 99% of their genetic material; they also estimate that the races of man are 50 times closer than are man and chimpanzee. Individuals whom sociobiologists consider unrelated share, in fact, more than 99% of their genes. It would be easy to make a model in which the structure and physiology important in behavior are based on the shared 99% and in which behaviorally unimportant differences, such as hair form, are determined by the 1%. The point is that genetics actually supports the beliefs of the social sciences, not the calculations of the sociobiologists. The genetic basis for the behaviorally important common biology of the species (in evolutionary order—bipedal walking, hand skills, intelligence) lies in the 99% and is undoubtedly modified by genes in the 1%.

The unity of mankind is in the genetic material, in the biology it determines, and in the basic human behaviors based on the biology. The sociobiological calculus is necessarily racist because geographical distance was a major factor in determining the formation of races (Barash, 1977b:311). In general, the further that two populations were apart at the time races were forming, the greater the genetic difference; hence, the less ethical responsibility people should have for members of the other group. The contrary argument is that most genes are shared and that gene frequencies in which races differ are behaviorally unimportant (with some exceptions such as sickle-cell disease).

As I stated at the beginning of this article, I am not concerned with proving that the position I am taking is correct, but rather with showing that the application of the genetic theory of natural selection requires research. The form of the study might be (1) evolutionary theory, (2) major research effort—where the time and energy must go, and (3) tentative conclusions. In applying evolutionary theory to human behaviors, sociobiologists make practically no effort to understand human behavior and they perform no research to validate their claims as far as human behavior is concerned. The personal opinions of the authors are presented as facts—or at least as serious contributions.

Postulating genes to account for behaviors is a major feature of the application of sociobiology to the interpretation of human behaviors. For example, in the last chapter of *Sociobiology* (Wilson,

1975), genes are postulated to account for more than 25 behavioral situations. There are conformer genes, genes for flexibility, genes predisposing to cultural differences, and many others. No evidence is given for any of these. Even the altruistic genes that are central to the whole concept of kin selection are not demonstrated for human beings. The logic is that there must be altruistic genes to account for altruistic acts—just as we learned many years ago that if there are criminal acts there must be criminal genes. There is, obviously, no need to postulate genes for altruism. It would be much more adaptive to have genes for intelligence, enabling one to be altruistic or selfish according to the needs of the moment.

Postulating genes to account for human behaviors allows sociobiologists to minimize the difference between the genetically determined biological base and what is learned. They are not worried by their repetition of the errors of the eugenicists, social Darwinists, or racists.

On the matter of homosexual behavior, Wilson states that it could be understood either by postulating selection for heterozygotes, or, following Trivers, by homosexuals helping relatives and thus their own genes (kin selection). This assumes that the actions described as homosexuals in our culture are general. It illustrates the repeated mistake of regarding our culture as synonymous with human nature. Consider highland New Guinea, where boys and men lived in men's houses and all males had homosexual relations. The women lived with the younger children and pigs, and husbands and wives had intercourse in the daytime by the gardens. Just as in the cases of polygamy mentioned earlier, the whole pattern of life was different from ours—and it is the patterns that must be compared. It is futile to take an American word and guess about possible genetic influences.

Just as the power of the theory of evolution in its 19th-century form gave people the confidence to arrange stone tools, animal behavior, or human customs into supposedly chronological sequences, so the power of the genetic theory of natural selection gives sociobiologists the confidence to provide genetic explanations for social customs without producing careful analyses or supporting facts.

Sociobiologists revel in the hope that genes will be found to account for the behaviors that social scientists think have already been explained by social facts. Fortunately, it is possible to accept the theory of evolution and the important of genetics without ignoring the known historical record, social science, or the psychology of the individual.

Summary

Recently there has been a great increase of interest in animal behaviors. It has been claimed by both popular writers and scientists that knowledge gained by the study of other animals may be directly applied to the interpretation of problems of human social behaviors. The proponents of this point of view minimize the difference between genetics and learning, attack Durkheim and the notion of social facts, and postulate genes as important in a very wide variety of human behaviors.

Surely the full understanding of the behaviors of any species must include biology. Biology is the basis for the differences between species. But the more learning is involved, the less there will be any simple relation between basic biology and behavior. Because of intelligence, speech, and recent history, human beings provide the extreme example in which highly diversified behaviors may be learned and executed by the same fundamental biology. Biology determines the need for food, but it does not determine the almost infinite number of ways in which this need may be met.

The desire, both popular and scientific, to apply biological evolutionary thinking to the interpretation of human behaviors has been, and is, so strong that conclusions came long before facts in social evolution, racism, and eugenics. Even the archeological record was misinterpreted by the belief that all that needed to be done was to arrange the stone tools in the order of simple to complex. History was even more distorted by arranging cultures into supposed evolutionary stages.

Sociobiologists are now repeating many of the errors of the past. *The laws of genetics are not the laws of learning,* and as long as sociobiologists confuse these radically different mechanisms, sociobiology will only obstruct the understanding of human social behaviors.

Positively, the increase of interest in animal behavior and the enthusiasm that has come with sociobiology will lead to the development of a biologically and socially based behavioral science. Such a science will be interdisciplinary, but will, in my opinion, contain a large element of social science and will be very different from the science suggested by Wilson (1975) in *Sociobiology*.

Negatively, as far as the interpretation of human behavior is concerned, if the application of biological thinking amounts to ignoring history, sociology, and comparative studies, and to postulating genes to account for all sorts of behaviors, it will amount to no more than a repetition of the errors of the past.

BIBLIOGRAPHY

Alcock, J. 1975 *Animal Behavior: An Evolutionary Approach*. Sunderland, Mass: Sinauer Associates.

Alexander, R. D. 1974 "The Evolution of Social Behavior." *Annual Review of Ecology and Systematics*, 5:325–83.

Altman, J. 1966 *Organic Foundations of Animal Behavior*. New York: Holt, Rinehart & Winston.

Ardrey, R. 1961 *African Genesis*. New York: Atheneum Press.

Barash, D. P. 1977a "The New Synthesis." *The Wilson Quarterly*, 1:108–20.

———— 1977b *Sociobiology and Behavior*. New York: Elsevier.

Bateson, P. P. G., and R. A. Hinde 1976 *Growing Points in Ethology*. Cambridge: Cambridge University Press.

Bienen, H. 1970 *Tanzania*. Princeton: Princeton University Press.

Blurton-Jones, N. 1975 "Ethology, Anthropology, and Childhood." *In*, R. Fox (ed.), *Biosocial Anthropology*. London: Malaby Press.

Brown, J. 1975 *The Evolution of Behavior*. New York: Norton.

Carpenter, C. R. 1934 "A Field Study of the Behavior and Social Relations of the Howling Monkeys (*Alouatta palliata*)." *Comparative Psychology Monographs*, 10:1–68.

———— 1940 "A Field Study in Siam of the Behavior and Social Relations of the Gibbon (*Hylobates lar*)." *Comparative Psychology Monographs*, 16:1–212.

Cohen, M. N. 1977 *The Food Crisis in Prehistory: Overpopulation and the Origins of Agriculture.* New Haven: Yale University Press.

Eibl-Eibesfeldt, I. 1975 *Ethology: The Biology of Behavior,* 2nd ed. New York: Holt, Rinehart & Winston.

Galkidas-Brindamour, B. 1977 "New Insights about the Behavior of Indonesian Orangutans." *L.S.B. Leakey Foundation News,* Spring/Summer, p. 4.

Goodall, J. 1968 "The Behavior of Free-Living Chimpanzees in the Gombe Stream Reserve." *Animal Behaviour Monographs,* 1:161–311.

——— 1977 "Watching, Watching, Watching." *The New York Times,* September 15, 1977, p. 27.

Goodall, J., and D. A. Hamburg 1975 "Chimpanzee Behavior as a Model for the Behavior of Early Man: New Evidence on Possible Origins of Human Behavior." *American Handbook of Psychiatry,* 2nd ed. New York: Basic Books.

Goodman, M., and R. Tashian 1976 *Molecular Anthropology.* New York: Plenum Press.

Hamburg, D. A. 1963 "Emotions in the perspective of human evolution." *In,* P. Knapp (ed.), *Expression of the Emotions in Man.* New York: International Universities Press.

Hamilton, W. D. 1975 "Innate Social Aptitudes of Man: An Approach from Evolutionary Genetics. *In,* R. Fox (ed.), *Biosocial Anthropology.* London: Malaby Press.

Hinde, R. A. 1974 *Biological Bases of Human Social Behavior.* New York: McGraw-Hill.

Isaac, G. L., J. W. K. Harris, and D. Crader 1976 "Archeological Evidence from the Koobi Fora Formation." *In,* Y. Coppens, F. C. Howell, G. L. Isaac, and R. E. F. Leakey (eds.), *Earliest Man and Environments in the Lake Rudolf Basin.* Chicago: University of Chicago Press.

Kellogg, W. N. 1968 "Communication and Language in the Home-Raised Chimpanzee. *Science, 162:*423–27.

King, M. C., and A. C. Wilson 1975 "Evolution at Two Levels in Humans and Chimpanzees. *Science, 188:*107–16.

Lancaster, J. B. 1975 *Primate Behavior and the Emergence of Human Culture* (Basic Anthropology Unit Series). New York: Holt, Rinehart & Winston.

Lenski, G., and J. Lenski 1974 *Human Societies: An Introduction to Macrosociology,* 2nd ed. New York: McGraw-Hill.

Lindburg, D. G. 1977 "Feeding Behavior and Diet of Rhesus Monkeys (*Macaca mulatta*) in a Siwalik forest in North India." *In,*

T. H. Clutton-Brock (ed.), *Primate Ecology.* New York: Academic Press.

Lorenz, K. 1963 *On Aggression.* New York: Harcourt, Brace & World.

MacLean, P. 1970 "The Triune Brain, Emotion, and Scientific Bias." *In,* F. O. Schmitt (ed.), *The Neurosciences: Second Study Program.* New York: Rockefeller University Press.

Mayr, E. 1972 "The Nature of the Darwinian Revolution." *Science,* 176:981–89.

Medawar, P. B., and J. S. Medawar 1977 *The Life Science—Current Ideas of Biology.* New York: Harper & Row.

Morris, D. 1967 *The Naked Ape: A Zoologist's Study of the Human Animal.* New York: McGraw-Hill.

Morris, S. 1977 "The New Science of Genetic Self-Interest." *Psychology Today,* February 1977, pp. 42–51; 84–88.

Myers, R. E. 1968 Neurology of Social Communication in Primates. *In,* H. Hofer (ed.), *Neurology, Physiology, and Infectious Diseases.* Proceedings of the Second International Congress of Primatology (Vol. 3). Basel, Switzerland: Karger.

——— 1976 "Comparative Neurology of Vocalization and Speech: Proof of a Dichotomy." *In,* S. R. Harnad, H. D. Steklis, and J. Lancaster, (eds.), *Origins and Evolution of Language and Speech.* Annals of the New York Academy of Sciences.

Nadler, R. 1977 "Sexual Behavior of the Chimpanzee in Relation to the Gorilla and Orang-utan. *In,* G. H. Bourne (ed.), *Progress in Ape Research.* New York: Academic Press.

Ploog, D. 1970 "Social Communication among Animals." *In,* F. O. Schmitt (ed.), *The Neurosciences: Second Study Program.* New York: Rockefeller University Press.

Robinson, B. W. 1967 "Vocalization Evoked from Forebrain in *Macaca mulatta.*" *Physiological Behavior,* 2:345–54.

——— 1976 "Limbic Influences on Human Speech." *In,* S. R. Harnard, H., D. Steklis, and J. Lancaster (eds.), *Origins and Evolution of Language and Speech.* Annals of the New York Academy of Sciences.

Roe, A., and Simpson, G. G. 1958 *Behavior and Evolution.* New Haven: Yale University Press.

Sahlins, M. 1976 *The Use and Abuse of Biology.* Ann Arbor: University of Michigan Press.

Schaller, G. B. 1963 *The Mountain Gorilla: Ecology and Behavior.* Chicago: University of Chicago Press.

Sociobiology Study Group of Science for the People 1976 "Sociobiology—Another Biological Determinism." *BioScience,* 26:182–90.

Struhsaker, T. T. 1967 "Auditory Communication among Vervet
Monkeys." *In,* S. A. Altmann (ed.), *Social Communication Among
Primates.* Chicago: University of Chicago Press.

——— 1975 *The Red Colobus Monkey.* Chicago: University of
Chicago Press.

Thorpe, W. H. 1972 "The Comparison of Vocal Communication in
Animals and Man." *In,* R. A. Hinde (ed.), *Non-verbal Communication.* Cambridge: Cambridge University Press.

Wade, N. 1976 "Sociobiology: Troubled Birth for New Discipline."
Science, 191:1151–55.

Washburn, S. L. 1959 "Speculations on the Interrelations of the
History of Tools and Biological Evolution." *In,* J. N. Spuhler (ed.),
The Evolution of Man's Capacity for Culture. Detroit, Mich.: Wayne
State University Press.

Washburn, S. L., D. A. Hamburg, and N. H. Bishop 1974 "Social
Adaptation in Nonhuman Primates." *In,* G. V. Coelho, D. A. Ham-
burg, and J. E. Adams (eds.), *Coping and Adaptation.* New York:
Basic Books.

"Why You Do What You Do; Sociobiology: A New Theory of Behavior." *Time,* August 1977, pp. 54–63.

Wilson, A., and R. C. Boelkins 1970 "Evidence for Seasonal Variation in Aggressive Behaviour by *Macaca mulatta.*" *Animal Behaviour,*
18:719–24.

Wilson, A. C., S. Carlson, and T. J. White 1977 "Biochemical Evolution." *Annual Review of Biochemistry,* 46:573–639.

Wilson, E. O. 1975 *Sociobiology: The New Synthesis.* Cambridge,
Mass.: Harvard University Press.

——— 1977a "Biology and the Social Sciences." *Daedalus,* 11:127–
40.

——— 1977b Preface. *In,* D. P. Barash (ed.), *Sociobiology and Behavior.* New York: Elsevier.

Wind, J. 1976 "Human Drowning: Phylogenetic Origin." *Journal of
Human Evolution,* 5:349–63.

Yerkes, R. M., and A. W. Yerkes 1929 *The Great Apes.* New
Haven: Yale University Press.

SOCIOBIOLOGY AND HUMAN NATURE: A POSTPANGLOSSIAN VISION

Human nature is too big, too fascinating, and too important a subject for major thinkers to resist. It is also too complex to fit the manageable visions they construct. Mene, mene, tekel, upharsin; one by one, these visions are weighed and found wanting—defeated either by the diversity of human behavior or by our inability to step outside our own class and culture; thus we promote socially constrained systems as deduced, universal truth.

E. O. Wilson's "sociobiological" vision is grounded in his reading of evolutionary theory and in his conviction that natural selection has built our behavior with almost the same intensity and directness that it invested in our bodily form. "Human sociobiology" has become the focus of a heated and confusing debate. As I have been among its critics, I should specify what I find so troubling. "Sociobiology" is not a general name for any evolutionary theory of behavior—for how could an evolutionary biologist deny that Darwinian processes can work on behavior as well as form? It is not merely a claim that biology will abet our Socratic search to know ourselves, or even that genes have something to do with human uniqueness—for I would accept both these statements as virtual *a prioris* of my profession. It is, for Wilson and his supporters, a specific claim based on a notion that I reject about the potency of natural selection. (From the late

From *Human Nature,* Vol. 1, No. 10 (October 1978).

1930s until the mid 1960s, panselectionism dominated Western evolutionary thought. Natural selection was declared virtually omnipotent in fashioning every bit of an organism's form, physiology, and behavior. A new pluralism is now sweeping through evolutionary theory, but sociobiologists have chosen to toe the old line.) Wilsonian sociobiology holds that natural selection, working on genetic variation, has molded our most distinct and general behaviors. These behaviors are (or were when they first evolved) optimal solutions to the Darwinian imperative that each individual struggles to maximize its genetic contribution to future generations. This imperative elicits a range of different, superficially contradictory, behaviors. Usually we do best by looking out for ourselves, but sometimes we do more for our genes by sacrificing our own bodies to ensure the survival of enough close kin; both selfishness and altruism have a Darwinian edge in the right circumstances. Darwinism is a theory of *genetic* change (all else, from cultural to technological "evolution," is analogy); genetic arguments about the adaptive nature of specific behaviors are central, not incidental, to human sociobiology.

Dr. Pangloss is reborn in human sociobiology. The adaptationist paradigm is his agent. We all know Pangloss's refrain about the "best of all possible worlds," but we tend to forget that his world is not a pleasant one, to say the least. It is simply the best that could arise under prevailing constraints. So too with human behavior on the sociobiological model. Men wage war and dominate women because the Darwinian imperative specifies it as optimally adaptive for individual reproductive success in our original ecological niche (still occupied today by the few remaining hunter-gatherers). Life may have been nasty, brutish and short, but nature specified no better.

And so we lived, in Panglossian optimality, for 99 percent of our tenure on earth—until we invented agriculture a mere 10,000 years ago and unleashed our precipitous cultural climb to the big apple and the neutron bomb. But changing culture did not subjugate or supersede the genetic nature that evolved for our life as hunter-gatherers; culture built upon it. Culture is the hypertrophy of the human biogram: "The genes hold culture on a leash. The leash is very long, but inevitably values will be constrained in accordance with their effects on the human gene pool."* "Human nature is, moreover, a hodge-podge of special genetic adaptations to an environment largely vanished, the world of the Ice-Age hunter-gatherer. Modern life, as rich and rapidly changing as it appears to those caught in it, is nevertheless only a mosaic of cultural hypertrophies of the archaic behavioral adaptations." Our modern world is Postpanglossian—the innate aggression that once led us to genetically advantageous warfare may now destroy us. This is our true human bondage; sociobiology will teach future leaders how to deflate such dangerous hypertrophies, and it will protect us by discovering what society may not manipulate because it lies too deep within our souls.

The debate on human sociobiology has focused properly upon the issue of genetic constraint. The political implications of biological determinism, buttressed by the sad historical record of its previous usages, have dominated the discussion. Quite apart from the intent of scientists who propose deterministic theories, unsupported claims for genetic coding of behavior lead to the defense or toleration of existing social inequalities as natural. Often lost in the heat of this argument has been the methodological point—that these claims are *unsupported*. Critics have never rejected sociobiology simply because they dislike a potential social message. We live with all manner of unpleasant biological truths, none more pressing and inevitable than death, and we will live graciously with human sociobiology if its premises be valid. I believe that the methodological flaws in human sociobiology are serious enough to incapacitate its central claim of strong genetic constraint imposed by natural selection for specific, adaptive behaviors. In *On Human Nature,* Wilson often approaches the political question with a commendable caution missing from his previous work. In so doing, he seems to think that he has adequately met his critics. Yet the foundation of our unhappiness has always been the methodological issue: human sociobiology is unsupported, not merely bedeviled by unfortunate implications. We may have been more sensitive to the flaws because we disliked the implications; but we didn't make

* This and all other quotations are taken from *On Human Nature* by Edward O. Wilson.

286 STEPHEN JAY GOULD

them up. *On Human Nature* exposes the flaws more baldly than Wilson's earlier work because it is more speculative in intent and because it deals exclusively with the subject we know least about—ourselves.

In the absence of direct evidence for genetic control, Wilson relies upon a large set of inferences, all problematical. "The question of interest," he proclaims, "is no longer whether human social behavior is genetically determined; it is to what extent. The accumulated evidence for a large hereditary component is more detailed and compelling than most persons, including even geneticists, realize. I will go further: it already is decisive." Wilson's first inference is based on universality: cultural invariants must be programmed. He presents a list of pervasive traits "as true to the human type, say, as wing tessellation is to a fritillary butterfly." He includes calendars, dream interpretation, surgery, visiting, weaving, and weather control. Later, he searches for invariants in the emergence of civilization and finds, for example, increasing division of labor, inequality of wealth, expansion of bureaucracies, and perfection of agriculture and irrigation. But what does this say about genetic constraint? Some behaviors are invariant simply because there are no alternatives, given our capacity to think in the first place (a genetic character to be sure, but Wilson has something far more specific in mind). Can you imagine extensive agriculture without an ability to reckon the seasons, people who never wonder what something so fascinating as a dream might mean, complex societies with declining bureaucracies, or an increasingly dense population unconcerned with improving its yield of crops?

Wilson's second inference relies upon analogy. He recounts a series of severe behavioral pathologies with an undoubted genetic foundation and asks us, on this account, to attribute variation within the normal range to similar causes. We may as well assume that all overweight people cannot help it because some very obese people do not metabolize properly due to a genetic disorder. Wilson points to the most severely retarded people as evidence for a base-level animality in all of us, arguing from a discredited model of the brain that these unfortunate people do not have a functional outer cortex and work only with the prehuman limbic system enclosed within. He even resuscitates the XYY theory of male criminality, stating

that "more specific forms of predisposition toward a crimal personality" may be coded by the extra Y, even though he admits that
existing studies discredit the idea.

Wilson's third, and most important, inference embodies his
favorite argument: if adaptive, then genetic—i.e., he presents
evidence that behaviors are adaptive and then proceeds, on this
account alone, to infer a genetic base for them. Wilson is aware
that adaptive behavior can arise without genetic selection, but he
promises to sort out the two mechanisms: "If human beings were
endowed with nothing but the most elementary drives to survive
and to reproduce, together with a capacity for culture, they would
still learn many forms of social behavior that increase their biological fitness. But as I will show, there is a limit to the amount of
this cultural mimicry, and methods exist by which it can be distinguished from the more structured forms of biological adaptation."

Yet I find no fulfillment of this promise. Most examples do not
go beyond the demonstration of adaptiveness. Two full chapters, on
altruism and religion, do little more than display the survival
value of these two perplexing phenomena. The chapters are full of
insight, but they do not buttress genetic claims. No genes need en-
code cooperation or conformity. Societies without cohesion may not
survive, even if their members have the same genetic make-up as
others who learned to cooperate and transmitted these beneficial
habits culturally. Cooperation and conformity are within the
capacity of the human genome; they need not be coded as specific
adaptations.

In an interesting discussion, for example, Wilson demonstrates
that several Indian peoples meet the predictions of an ecological
theory called "optimal foraging strategy." They share at low population densities in areas of poor and unpredictable resources; they
defend and establish territories where land is fertile and game
abundant. Wilson then invokes his concept of hypertrophy to claim
that "the biological formula of territorialism translates easily into
the rituals of modern property ownership." As part of this hidden
biology, he cites certain behaviors noted around vacation residences
in Seattle: visitors identify themselves and greet owners before
entering another property; children are asked where their parents
are; entrance to a house is preceded by an identifying noise,

a knock or ring if the door is closed, a vocal call if it is open; adults are held responsible for the territorial transgressions of children; identifying rituals are less formal in entering outdoor lots than houses. Now I finally understand why my neighbors always accosted my parents when I hit an inside-the-park homer (losing the ball through a broken second- or third-story window on my stickball court).

Many of Wilson's claims are more dubious because he relies so heavily upon a speculative tradition, unfortunately followed widely within evolutionary biology—the just-so story. Since he assumes that natural selection is nearly totipotent, it becomes legitimate to invent a story about its action when faced with an adaptive behavior. But suppose the behavior is not an adaptation at all? suppose it is "for" something else? suppose it is not a product of genetic natural selection? We cannot test these alternatives because genetic stories can always be constructed—and Wilson uses the possibility of construction itself as a criterion of validity. In many cases, he tells a story that precludes disproof. He wishes to argue, for example, that "carnivorism remained a basic dietary impulse," despite "cultural after effects that varied according to the special conditions of the environment in which the society evolved." He presents as established the idea that cannibalism due to chronic protein shortage was the hidden cause of Aztec human sacrifice. Yet he only mentions in an end-note the potent criticism that, to my mind, has virtually disproved the idea (see B. R. Ortiz de Montellano in *Science,* May 12, 1978). Agriculture and food tribute provided enough protein; only the well-fed elite were allowed to partake; and sacrifice coincided with times of plenty in the harvest, not with times of scarcity. He then attributes Indian veneration of cattle to the same dietary impulse: The Aryan invaders of the Gangetic Plain were confirmed meat eaters, but as populations grew denser after 600 B.C., meat grew scarcer and people struggled to conserve enough livestock for milk, dung for fuel, transport, and agriculture. In India, the poor people won, and the cow was reclassified as a sacred animal. But if inherent carnivory can explain everything from vegetarianism to cannibalism, then it is an unbeatable theory, and theories that cannot lose are not much use.

Wilson's perspective lies within the philosophical tradition of reductionism. He sees the sciences as a linear array, from physics at the bottom to sociology on top. Each rung, the "antidiscipline" of one above, seeks to explain the higher step in its terms. Darwinian sociobiology is the antidiscipline of the human sciences. The rationale of cultural practices should be sought in biological principles; the apparently autonomous actions of human groups should be reduced to the Darwinian advantages of individuals.

Wilson's search for biological roots and individual advantage as the cause of social phenomena reflects a selective myopia in his interpretation of history and his recommendations for our future. He views 17th-century witch mania as "rooted in the self-seeking of individuals." It provided psychological protection as a substitute for the "positive witchcraft" of previous catholicism. Accused witches were often poor women who had sought favors from their accusers and been refused. If disaster then struck the accuser, he could fasten blame on the "witch" and rationalize his biologically selfish behavior thereby. But Wilson misses the entire dimension of social utility to groups in power. Witch hunting was a state institution, with its own bureaucracy, manuals of procedure, and manufacturies for instruments of torture. Its social role in the suppression of dissent, especially in the face of the counter-reformation, cannot be denied. It did not just placate individual psyches. Similarly, Wilson would secure world peace by overcoming the biology of territory and tribalism. We have a biological tendency to specify ingroup and outgroup, and to regard outgroups as less than human and therefore subject to legitimate elimination. So we must "create a confusion of cross-binding loyalties" to enlarge the ingroup, hopefully to all of us. This would be a fine thing, but we might also consider the economic roots of war: poverty, oppression, and exploitation of resources. Wilson misunderstands and caricatures Marxism as if this rich tradition of thought encompassed nothing more than the most deterministic version of Marx's theory of historical stages. He might seek some guidance from Marxists and other economic historians in expanding his realm of causation to include the economic motivations of groups and the dynamics of social classes operating in their own interests. Groups do have an autonomy of motive and action that transcends the genetic advantages of their members. This very autonomy precludes reduction to biology. I do not indulge, as Wilson might claim, in obscurantism or mystification. I simply claim that the world is ordered hierarchically and that explanations must be sought at appropriate levels.

The critics of sociobiology do not seek to deny the importance of constitutional factors in human nature. No search could be more interesting or important. I believe that Wilson has made a fundamental error in identifying the wrong level for biological input. He looks to specific behaviors and their genetic advantages, and invokes natural selection for each item. He tries to explain each manifestation, rather than the underlying ground that permits their manifestation as one mode of behavior among many—as if each great classical symphony were its own best solution, rather than the varied product of some loosely constraining generating rules and a large dollop of individual genius. Thus, Wilson asks: "Are human beings innately aggressive? This is a favorite question of college seminars and cocktail party conversations, and one that raises emotion in political ideologues of all stripes. The answer to it is yes." As evidence, Wilson cites the prevalence of warfare throughout history. "The most peaceable tribes of today," he writes, "were often the ravagers of yesteryear and will probably again produce soldiers and murderers in the future." But if some peoples are peaceful now, then aggression itself is not encoded, only the *potential* for it. If "innate" only means possible, or even likely under common circumstances, then it cannot bear Wilson's claim that natural selection works to optimize the manifestations. We should seek constitutional factors in the generating rules, not in the epiphenomena.

MICHAEL A. SIMON

BIOLOGY, SOCIOBIOLOGY, AND THE UNDERSTANDING OF HUMAN SOCIAL BEHAVIOR

The human being is an organism. Its behavior is a kind of animal behavior. Since the behavioral tendencies of animals, like their anatomy and physiology, are explainable as the results of processes of biological evolution, it would seem reasonable to suppose that the same is true of human behavior. Many animals, furthermore, are social. If we can look to biology for the explanation of animal behavior in general, we should be able to do so for their social behavior. To be concerned with what beings do to each other, for each other, and with each other, is necessarily to be concerned with biological phenomena.

If a biological approach is to be useful with respect to explaining social behavior, that behavior, whether it is engaged in by humans or by (other) animals, must be representable as the result of biological forces. Specifically, anything that counts as social behavior will have to be explained in terms of a creature's inherited biological nature and its environmental adaptation. What makes an account biological and thereby distinguishes it from other environmental accounts is that the adaptive behavior it describes is presented as having been determined by the organism's biological makeup. A biological explanation must appeal to features that evolution may be presumed to have fixed in the creature's genotype.

The task of this essay is to examine some of the implications of a biological approach to human social behavior. The kinds of phenomena whose explanation I shall be concerned with consist of human actions plus certain events and states of affairs that are the results of actions. These phenomena include social practices, institutions, and routinized patterns of voluntary behavior such as are fixed by rules, roles, and relations, in addition to all of the kinds of individual and group interaction that occur within a society. They do not include reflex movements or other bodily processes not subject to voluntary control. Nor do they include capacities, which are not actions themselves but only fix the limits of action. The capacity to learn to speak a language or write a sentence is not a social phenomenon, though the delivering of a sermon or the writing of a letter is. It is not, therefore, an explanation of a social action or practice to indicate how a particular structure or capacity has come into being, even though no performance of the action or exercise of the practice would be possible without that structure or capacity.

It is obvious that not all animal species behave alike; if they did, it would not be necessary to study the behavior of more than a single species in order to determine the way all species behave—including humans, presumably. Ethology, which is the study of animal behavior in zoological contexts, is based on a recognition of the need to take this fact seriously. To the early ethologists, the study of animal behavior meant the study of instinct. They concentrated on behavior that could be seen to be rigidly stereotyped, such as nest-building among insects and mating dances among tropical fish. While the methods were easily extended to birds and mammals, including the higher primates, the paradigm remained the same: Behavior that is uniform throughout the species is triggered off by specific environmental cues and is stable under otherwise changed environmental conditions. The project was to identify and explain the things that animals do as a result of their inherent natures.

Later ethologists, more aware of the fact that animals do not behave in their native environments in all of the same ways that they do under controlled and restricted conditions, have construed the study of animal behavior more broadly and have allowed it to include the study of learning and other environmentally induced modifications of behavior. There is a tendency among modern students of animal behavior to avoid the use of expressions like "instinct" and "innateness". Behavior, they point out, is always a joint product, a result of interaction between a genotype and an environment, and the role of hereditary factors is a matter of degree. And since animals possessing a given species-inheritance live in a variety of environments and in fact exhibit different behavior depending on the features of the ecological niche they occupy, the study of their behavior should not be restricted to what can be identified as instinctive. If, as often seems to be the case, there is no inherent structure that rigidly and specifically determines a unique type of behavior, independent of the nature of the environment, then it is at least misleading to say that the behavior is programmed by heredity. The biology of behavioral adaptation turns out to be a good deal more complicated than is suggested by an approach that concentrates on trying to find out which behavior is instinctive or innately determined.

Nevertheless, what reliance on an ethological perspective typically signifies when human behavior is being considered is that an effort is being made to show that human biological inheritance must be invoked to explain why people and societies behave as they do. Thus the point of Lorenz's attempts to extrapolate to humans the results of his studies of aggression in animals was to show that humans have an instinctive aggressive drive that leads them to kill members of their own species. Similarly, the principal thrust of the recent attempts to offer biological accounts of phenomena such as human altruism, xenophobia, the sexual division of labor, and homosexuality has been to suggest that it is the genetic structure of human populations, not environmental influences, that is responsible for the patterns of social behavior that we observe. Instead of trying to reveal the way biological structures and environmental conditions interact to produce human behavior, the aim seems rather to demonstrate its dependence on the species' genetic constitution. Examination of human behavior from a biological perspective has been thought to be capable of helping us discover human biological nature.

If there is such a thing as human nature and biology can be used to investigate it, the biologist is in an important position for helping us discover why human societies are the way they are. If we could find out what is natural for the human species, what people are really like, independent of cultural influences, we would know what is fixed by biology and what needs to be explained as a result of our environment. Such knowledge could serve as a foundation for social science. For a conception of human nature that is grounded in a knowledge of biologically inherent tendencies would provide a base for sociological and psychological explanations; it would constitute a background against which the things people do, including those that are environmentally influenced, can be displayed. Just as in Newtonian mechanics the concept of motion at constant velocity (of which rest is merely a special case) yields a standard with reference to which all other types of motion are explained, so a genetic or ontogenetic concept of human nature would enable us to know which social phenomena we ought to seek explanations for, and which, because they flow directly from human biological nature, we ought to regard as intelligible in themselves.

A science that can establish which human social patterns will remain invariant through major cultural change, and which patterns can be modified, could have considerable social importance. If we know how deeply—or how shallowly—rooted human aggressive or territorial behavior is, for instance, we would have a better idea of the significance of social institutions as means of maintaining peace. It could be very useful to know whether or not there is a genetic basis for competitiveness, whether xenophobia is a universal heritable characteristic that will arise in every society, and whether moral commitment is entirely learned or is subject to genetic programming. If we could discover the human species' natural propensities, we might then have better insight into why various social practices are adopted and what sorts of alternatives are possible.

A science that has been heralded as holding the promise of satisfying these aims is sociobiology. This relatively new discipline, whose practitioners look upon ethology merely as an adjunct, is supposed to offer "the systematic study of the biological basis of all social behavior."[1] Its program is to investigate animal populations, using not only the methods of comparative ethology but also the conceptual frameworks of genetics, including population genetics, and ecology.

What have the ethologists and sociobiologists achieved thus far with respect to gaining an adequate biological conception of human nature? In fact, the results have been extremely meager; at best, they are highly controversial. The problem is not that animal behavior has not been adequately investigated, but rather that what has been discovered has not shed much light on humans. Konrad Lorenz's studies of geese, wolves, and rats, for example, have definitely not succeeded in demonstrating the presence of an instinctive aggressive drive in humans, nor have the studies by George Schaller on gorillas and Jane Goodall on chimpanzees, indicating a general lack of aggression in higher primates, proven that humans are not innately aggressive. Sociobiologists have speculated that behavior such as human altruism, social conformity, and even creativity and entrepreneurship is controlled by genetic factors, but there appears not to be any direct scientific evidence that this is the case.

In order to grasp the reasons for the lack of success of the project of actually discovering a genetic or biological basis for human social behavior, we need to consider what one would have to do to establish that a human behavioral trait is based on species inheritance. First, one would have to become convinced that the trait is universal. A way of finding out whether a trait is universal is to make cross-cultural comparisons and determine that there are no counter-instances within the species. But would this show that the trait is genetically based, rather than a result of common environmental influences? E. O. Wilson, in his book, *Sociobiology*, suggests that it would, especially when the trait is found in all, or nearly all, other primates.[2] Although he allows for the possibility that some traits that are present throughout the rest of the primates "might nevertheless have changed [i.e., mutated] during the origin of man," he does not seem to consider seriously the possibility that they may be environmentally induced in humans. The fact that Wilson has said of qualities that are "distinctively ineluctably human," that "they can be safely classified as genetically based,"[3] indicates that he simply assumes that whatever is universal in humans must be fixed in their common genotype. Another possibility, not ruled out by the evidence, is that what is most distinctive in humans, beyond their "distinctively ineluctably human" traits, is their adaptability, their capacity to learn what in other animals is already programmed into the genes. Humans have no instinct to eat only edible food or to drink only potable water, but must acquire these tendencies on an individual basis in order to survive. It is very likely true that whatever is universal in humans is biologically significant, in the sense of contributing to the perpetuation of the species, but it does not follow that any of these qualities must be determined genetically.

One of the most striking features of social behavior in humans and certain other animals is its susceptibility to modification. Almost any kind of social behavior can be inhibited or provoked by means of sufficiently drastic manipulation of the creature's environment, often well within the limits of viability. Dogs can be domesticated, ordinarily peaceful monkeys can be trained in aggressiveness, and birds can be gotten to ignore their young. Behavior that can plausibly be designated as innate or instinctive must be stable throughout a range of environments. Whether or not a type of social behavior can be said to be determined by a creature's biological nature will therefore depend on whether the conditions under which the behavior fails to manifest itself can be classified as deviant or unusual.

We cannot justly characterize as instinctive or as rooted in human nature any kind of social behavior for which exceptions are known, unless we are able to explain away the exceptions as results of abnormal or unnatural conditions. We can justify calling eating instinctive, but only because we recognize as extraordinary the conditions under which a person will voluntarily starve to death. We could not, on the other hand, in the face of evidence indicating the existence of nonaggressive tribes, reasonably maintain that human aggression is innate; to do so would be a form of ethnocentrism. There is no conceivable evidence that would decide between the hypothesis that aggression is a basic drive that is sometimes masked or repressed, and the hypothesis that it is something that is produced only when other drives are thwarted and/or when certain types of stimuli are present. So far as the way people act is variable across cultural lines, there is no discoverable human nature that can be said to determine human social behavior.

If is it possible to investigate human nature, the only way that this can be done is to concentrate on aspects of human behavior that either do not vary at all or vary only in ways that can uncontroversially be exhibited as results of bizarre conditions. There are, it happens, a number of cross-cultural universals that have been at least tentatively identified. Male-male competition, for example, is something that has not been specified as absent in any culture that has been studied. Another universal seems to be the avoidance of incest by taboo. Others that have been mentioned include sexual inhibition and shyness, play, male dominance, and territoriality.

There is no evidence that these characteristics are genetically based, but neither is there any evidence that they are not. Since it is not clear what *would* show such universal features to be part of a species' genetic inheritance, the dispute is essentially a philosophical one. Where the traits in question fail to exhibit continuity with traits of other species in both structure and function, it is impossible to demonstrate homology—commonality of phylogenetic descent. As a consequence, there can be no convincing argument that the traits are products of Darwinian evolution. What is clear, in any case, is that the present defenders of the genetic interpretation have not presented evidence sufficient to justify their claims.

Let us assume, nevertheless, perhaps on the grounds that no alternative assumption is better supported, that behavioral universals such as the ones that have been suggested are genetically based and that they do reveal certain characteristics of human nature. I would like to argue that, even if that assumption is made, the generalizations that might be yielded would still not be of much help in understanding why people and societies behave as they do. One problem is that the principles by which we represent the patterns we observe as variations on common themes are likely to be true only at a level of abstraction that renders them trivial or common-place. Human social behavior does not consist of rigidly determined, fixed patterns that are uniform throughout the species; it rather varies considerably from one social group to another. Social science is typically interested in explaining these variations. The kinds of phenomena that we expect a social theory to explain include items such as racial prejudice, juvenile delinquency, war, individual and group rebellion, and social control. Whether or not these can all be seen as manifestations of biological drives or tendencies, it is still the differences among them as social phenomena that interest us. The enunciation of underlying biological generalizations, even if this is possible, will not be of much use. If the determinants of the specific differences among a widely disparate range of behaviors are not biological but sociocultural, it will not explain any of them to subsume them all under a common biological rubric.

As an illustration, let us consider the notion of human territoriality. If we observe the variety of ways in which humans maintain more-or-less exclusive occupancy of an area, we realize that territoriality, at least in humans, is not a matter of turning away unwanted visitors by means of a pattern of signals and responses common to the whole species. The rules regarding property and land-use are culturally determined and vary widely among different societies. The phenomena that a social science undertakes to explain include the variety of forms that human concern for property and territory takes. But if the concept of territoriality that sociobiology employs is made so broad as to allow senses as disparate as the ones that apply, respectively, to capitalist societies and to nomadic tribes, then it is not clear that much is explained by citing the fact of territoriality. It is doubtful whether anything substantive would be said if one were to assert that the human species is territorial, if this concept is made to fit—as at least one sociobiologist wants it to—not only societies that contain legal institutions that govern private property of all sorts, but also societies that do not allow exclusive use by tribes or families of any resource except the richest sources of vegetable foods, and even societies that under some conditions exhibit no territorial behavior at all.[4] Because the thesis that territoriality is a general human trait fails to signify a single unitary feature of either individuals or groups, it lacks explanatory force with respect to any of the multifarious behaviors that have been called territorial.

In order for a cross-cultural generalization to explain an observed practice or pattern of behavior, it must not be as loose or abstract as the sort we have been considering. On the other hand, even when the generalization is at the level of the phenomena to be explained, it will not go very far as an explanation of why people and societies exhibit the traits they do. The discovery that something is a universal human trait will show only that it is a general feature of human existence, that it is not merely the result of particular events and circumstances. It remains to be shown that the feature is a biological one, that it derives from the creatures' genetic inheritance.

Furthermore, in order for a generalization to be used to provide a biological explanation of the trait in question, it must be possible to show *how* certain biological characteristics are responsible for the behavior. If that can be done, however, then the argument that the behavior is biologically determined no longer depends on its being universal. The problem of showing how a creature's biological inheritance can account for a behavioral trait is the same whether the trait is universal or variable. In fact, we find that defenders of the biological approach do treat at least some of the variation that is found among human societies as due to genetic differences. Thus Wilson has suggested that the "genes . . . maintain a certain amount of influence in at least the behavioral qualities that underlie variations between cultures," and that "we can heuristically conjecture that the traits proven to be labile [i.e., traits that shift from species to species or genus to genus] are also the ones most likely to differ from one human society to another on the basis of genetic differences."[5] In other words, any behavorial trait, whether it is common to all societies or is a feature of only some societies, can be seen as revealing the genetic basis of human behavior. The gene hypothesis at this stage appears to be capable of explaining any distribution of traits whatever—and thus it explains nothing.

In order for a human trait to be explained biologically, it must first be "biologized," i.e., represented in terms that befit the study of organisms *qua* organisms. The problem with such biological reduction is that it is likely to sacrifice precisely those features of human social behavior that give it a socially or philosophically distinctive character. A case in point is the treatment of altruism. When Wilson defines an act of altruism as one that occurs "when a person or animal increases the fitness of another at the expense of his own fitness," he is specifically ignoring the distinction between acts that are performed with the *intention* of benefiting others and acts that merely *turn out* to have this effect. Using this definition, Wilson is able to label as cases of altruism, not only the behavior of dolphins in cooperating to rescue their wounded but also the labors of sexually neuter workers among the social insects, the warning calls of small birds, the defense of a colony by the soldier caste of termites, and even the behavior of bees that lose both their stings and their lives when attacking a predator, thereby leaving a

chemical deposit that serves as a signal to summon additional defenders. (It is interesting that Wilson chooses to characterize this
type of apian behavior as "kamikaze attacks."[6] One would be unlikely, I think, to call an *actual* kamikaze attack a case of altruism
unless the mission were a *voluntary* one.) By disregarding what
makes the creature do the beneficial acts it performs—its reasons,
if you will—the sociobiologist is likely to miss the entire point of
designating a piece of behavior as altruistic.

A sociobiologist might perhaps reply that he is not interested in
what the *point* is of calling a type of behavior altruistic; rather he
is concerned to offer an explanation of how the behavior we call
altruistic arises and why it persists. Indeed, what Wilson and others
have tried to show is merely that such behavior, whether it occurs
in humans or in dolphins or in the social insects, serves to help as-
sure the transmission of a species' genetic material. The fact that
Wilson is willing to concede that, as far as human altruism is concerned, the specific form of the acts performed is to a large extent
culturally determined does not at all prevent him from insisting
that the impulses that give rise to these acts are themselves the products of evolution through the genes.[7]

Nevertheless, I think it can clearly be affirmed that whatever is
explained by the sociobiologists' speculation, it is not altruism as
the concept is ordinarily understood, but rather something much
more general, of which human altruism may be viewed as a special
case. Not everything we do that benefits others more than it benefits ourselves is a case of altruism. And just as we would hardly be
willing to accredit a putative study of suicide that failed to distinguish it from accidental death, it is difficult to imagine anything but
confusion resulting from trying to explain the social phenomenon
of altruism without distinguishing it from the unintentional or incidental conferral of benefits. To the extent that understanding human institutions and human social interactions requires grasping
distinctions that have to do with reasons and intentions, and these
are part of the residue that biobehavioral reduction leaves behind,
the social behavior of humans will not be explainable biologically.

If there is any way that human biology can be said to influence
human behavior, other than by setting the limits that fix human
capacities, it must be through determination of a person's feelings

and inclinations. The idea is that what people do in any given situation depends on their natural urges, their inborn desires, and the tastes and preferences that are part of their biological makeup. All of these features are supposed to be fixed by the genotypes; the genotypes are the result of natural selection; and the natural selection is based on the adaptive advantages that the genotypes confer, or did confer at an earlier time, either on the individuals who perform the actions to which these genotypes lead or on populations in which these actions occur.

What the theory—if that is what it may be called—implies is that social practices that persist do so not because they are adaptive or because of the weight of cultural tradition but because of propensities that reside in the genes. The theory also seems to require that when people opt for what their tradition and their cultural inputs encourage them to do it is because of innate tendencies. This proposition is not only unsupported by evidence, but it is entirely gratuitous, given that other explanations, viz. cultural ones, are readily available. We do not need to invoke innate tendencies to account for racism, for example, especially when we already have available to us equally well-supported explanations in terms of social and economic factors. Explanations of social practices in terms of innate preferences and propensities are inadequate because they cannot cope adequately with exceptions: whenever we are confronted with a counter-instance, either we must suppose that the tendency is one that can be overridden, in which case the factor is too weak to provide a satisfactory explanation of the behavior, or else we must make an *ad hoc* postulation of a genetic difference in order to account for the exception.

The same considerations apply to attempts to provide a biological explanation of the apparently ubiquitous incest taboo. There is evidence, albeit indirect, that the avoidance of the dangers of excess homozygosity, viz., physical and mental defects, could very well be the basis of the taboo, but this, the giving of a biological reason, would not show that incest-avoidance has a genetic basis. There would have to be a biological mechanism, one that works through natural preferences to make individuals not *want* to make with kin. But it is doubtful whether such a mechanism is operative. For, as a number of people have pointed out, if incest avoidance were instinctive, incest would not have to be illegal. Some people do (knowingly) commit incest, and unless we are going to introduce the utterly wild assumption that these people are genetic variants, we must suppose that, whatever basis there is in human biological evolution for the fact that humans almost always do avoid incest, it is not enough to explain the taboo.

Wilson has suggested that the way a cultural tradition may become established is by means of social reinforcement of natural tendencies that have been selected for because of the adaptive advantages they confer. With regard to the incest taboo, a mechanism that has been supposed to operate involves what has been called "the precluding of bonds": kinship relationships such as between fathers and daughters, mothers and sons, and brothers and sisters seem to exclude the possibility of formation of other types of bonds.[8] The evidence that has been cited for this hypothesis includes studies in Israeli kibbutzim wherein it was found that, among unrelated members of the same kibbutz peer group who had been together since birth, there are no recorded instances of heterosexual activity, despite the absence of formal or informal pressure, and that all of the marriages that occurred were with persons outside the kibbutz. The inference drawn is that social prohibitions on incest may have arisen as a result of evolved natural inhibitions, inhibitions that have persisted because those who have them tend to leave a larger number of fertile offspring than do those who lack them.

The model is not an implausible one, and could be generalized to cover other sorts of cases as well. Cultural rules and social beliefs could very well have arisen as rationalizations of people's natural preferences. The difficulty with this suggestion, besides the fact that it lacks direct empirical support, is that these preferences, since they are known to be overridable, are again too weak to provide significant explanatory power. We know from comparative anthropology that human beings are capable of internalizing a number of quite different norms. Innate preferences of the sort invoked are clearly not strong enough to prevent some tribes from drawing the incest line between cross-cousin and parallel-cousin marriages, for example. It is very tempting to try to derive norms from natural tendencies; but it is impossible to say which norms *must* be so derived. A norm is something that governs *voluntary* behavior, and there is no norm that is not susceptible to replacement by a substitute.

As a final example, we may consider the matter of the sexual division of labor, whereby, as Wilson puts it, "women and children remain in the residential area while the men forage for game or its symbolic equivalent in the form of barter or money."[9] Wilson has suggested that the basis for this division of labor, which he takes as revealing a genetic bias, may lie in the facts that males are, on the average, demonstrably more aggressive than females from the beginnings of social play in infancy, and that they tend to show less verbal and greater mathematical ability. If these differences exist, they may go part of the way in explaining why some people dominate others, and also why certain professions have a disproportionate number of men in them. But they do not explain the institutionalization of male-female dominance patterns and the resulting role-distributions, nor do they explain the extent to which existing social patterns reveal polarizations that are as pronounced as they are. A basketball team may dominate another team whose members are on the average shorter than those of the first team, but only because each team is organized as a unity for the sake of demonstrating its collective dominance. You cannot derive a culture of male-female dominance, or deduce a strict or nearly strict sexual division of labor, merely from a set of statistical differences between males and females.

What the biologist who seeks a basis for understanding human social behavior in the study of nonhuman animals is trying to show is that some of what we know about animals is also true of human beings. Since human behavioral traits, unlike anatomical features, cannot, for the most part, be established as homologous, as based on common ancestry, the ethologist is forced to rely on analogy. But analogies that concern patterns of human social behavior and similar displays among animals are unconvincing. The behavior typically is either disanalogous or not known to be analogous just at the point where analogy is the most crucial: the way the behavior is mediated. If behavior that is rigidly determined by a mechanism whereby a specific releaser triggers off an internal response has, as its counterpart in human beings, behavior that is

subject to nothing like such causal determination, we are not licensed to infer anything whatever concerning the biological basis
of the behavior in question. It is also presumptuous to assume, in
the absence of any evidence pointing to a specific kind of mechanism, that a universal feature of human societies is a biologically
inherent characteristic of that species, regardless of how many other
species exhibit behavior that is analogous. When similar behavior
is observed across species lines but the biological mechanism shown
to operate in one species is known not to operate in the other, it
is not reasonable to assume that there must be some *other* biological mechanism present that accounts for the effect in the
second species. Human behavioral tendencies *could* be species
characteristics having significant genetic components, but they
could also have arisen in other ways; and since we know that the
behavior is in any case subject to environmental modification, biological conjectures are extremely unlikely to contribute to understanding why the behavior occurs.

A behavioral trait that is common to all human societies must be
shared by some other animals, all other animals, or no other animals. As far as behavior that is unique to humans is concerned, it
is clear that comparative studies with other species will not provide
much illumination, regardless of whether the human behavior is
throught to be genetically determined or not. Studying behavior
that is performed by some other species as well as by humans, or
even by all closely related species, on the other hand, is also not
going to yield results that can automatically be extended to humans.
If what is true of animal societies is also true of human societies,
this can be established only by studying human beings on their own.
We cannot *assume* that animals belonging to different species will
behave in similar ways under similar circumstances, nor can we assume that behavior that is common to two or more species and has
a genetic basis in one is equally heritable in the others. We simply
do not know what to make of our observations of animals. So far
as social behavior is concerned, there is not enough force in any
conclusion that could be reached regarding any or all nonhuman
species to give its extension to humans a significantly higher antecedent probability of being true than could be assigned to the
proposition that people are exceptions.

Because discoveries about the ways animals behave are often so *interesting,* it is tempting to believe that there must be something that we can learn about humans as a result of these studies. One can easily be impressed by seeing the way monkeys or wolves avoid destructive intraspecies fighting by means of dominance hierarchies, or by observing the effects of crowding on social harmony among rats and hippopotamuses. It is fascinating to discover that female baboons do not compete directly but rather vie with one another in terms of the position of the males associated with them in the dominance hierachy, or that rats, who will try to kill an intruder when it cannot escape, will not (unlike the Royal Canadian Mounted Police) pursue it into the wilds. By telling us a lot about animals that we did not know, animal behaviorists can, perhaps, help us to discover that we are more like animals than we thought we were.

It has been suggested that, although animal behavior studies do not prove anything concerning natural human tendencies, or even that there are any, they may provide a source of ideas for subsequent studies that could be carried out on humans.[10] Perhaps noticing that crowded conditions among cats give rise to the emergence of a despotic leader, or that young male African waterbucks, who are not tolerated by the older, territorial males, form roving "bachelor herds," can serve as a guide to looking for connections between elements within human societies. Biological studies may serve as means of generating important questions and of providing tentative answers that might then be examined as to their relevance for humans. We can ask, for example, how other animals avoid bloodshed, and then try to find out how ritualization serves this function. We can observe the way some species preserve social stability through adherence to dominance hierarchies that are rarely if ever challenged, and we can see what happens to previously peaceful animal colonies under conditions of higher population density. While no legitimate inferences may be drawn as to what results would be obtained were these ideas to be tested on humans, studies such as these do represent a considerable source of possible experiments to be carried out within the human domain.

Nothing that is a possible source of ideas deserves to be summarily dismissed, of course, and animal behavior studies may be a particularly rich source. They may also be no better a source of ideas about human behavior than could be afforded by travel to foreign lands or by reading imaginative fiction. The role of a source of inspiration is a very important one, but it is also very limited. I suspect that what makes animal societies a particularly attractive source is not so much the wealth of illustrations they afford as it is the unsupported notion that what is true of animals and can be applied to humans is likely to be true of them. But like the free play of fantasy, knowledge of the ways animals conduct themselves does not tell us anything about what happens in human life.

Citing ethological discoveries in the context of considering human social life has an effect that is largely rhetorical. Like Aesop's fables, facts about the ways animals behave are often thought to provide us with "lessons" as to how we ourselves might behave, quite apart from whether or not they reveal the way we do in fact behave. Specific findings with regard to animals, like stories or myths, neither establish nor refute assertions concerning what human behavior is or ought to be; the most they can do is serve to counter other claims that have been made based on other examples. Finding that higher primates lack intraspecies aggression would not lead to any reliable conclusion concerning humans, but might serve as an antidote to the claim, based on analogies between humans and certain animals that do fight, that human beings are innately given to fighting. The popularity of the sort of popular ethology that is associated with the names Konrad Lorenz, Desmond Morris, and Robert Ardrey is mainly due to these rhetorical effects. What these works offer are essentially anecdotal accounts of animal and human behavior enshrined in a general ethological perspective.

There is in these accounts a definite undercurrent that suggests that people do what they do, not for reasons or a result of conditions brought about by other people or by cultural influences, but because of internal forces that we have all inherited from our remote animal ancestors and which cannot easily be resisted. Thus Lorenz urges that "humanity must give up its self-conceit and accept that humility which is the prerequisite for recognizing the natural laws which govern the social behavior of men."[11] The reason "why reasonable beings . . . behave so unreasonably," he believes, is that human social behavior "is still subject to all the laws prevailing in all phylogenetically adapted instinctive behavior."[12] The lesson being taught is that we ought to resign ourselves to accepting a nonrational basis to our social behavior, and acknowledge the truth of such propositions as one that Ardrey affirms, that "man is a predator with a natural instinct to kill with a weapon."[13]

We have seen that none of these conclusions is warranted by the evidence ethologists have presented or are likely ever to turn up. Failure to realize how short the reach of their discoveries is has led some of these researchers to think they can derive socially significant results having a strong normative content—such as the inference that there is a basis in nature for private property or for an authoritarian family structure. Other students of animal behavior have been more circumspect and have tended to avoid such excesses, but they are not always willing to deny that their investigations *could* yield results of a prescriptive nature. Thus Wilson, while explicitly warning against the dangers of committing the naturalistic fallacy—the supposed mistake of trying to infer what ought to be from what is—nevertheless maintains, for example, that if the theory of "innate moral pluralism" is correct, then "no single set of moral standards can be applied to all human populations, let alone all sex-age classes within each population."[14] He also suggests that lessons can be derived, albeit not deductively, from discoveries such as those concerning the connection between crowding and aggressive behavior in cats and rats.[15] The important issue is not whether Wilson or any other sociobiologist has been completely faithful to the antinaturalistic stricture, however, but rather whether even the allegedly factual inferences regarding human beings are warranted. *Of course* facts are relevant for what ought to be done; the major question is whether the sorts of facts that sociobiology comes up with are ones that shed any light on the social behavior of humans.

Although I have been arguing that conclusions drawn from studying animal behavior concerning human societies lack adequate scientific basis, I do not wish to assert that ethologists and sociobiologists can never turn up significant insights about people. Scientific inference need not be regarded as the sole vehicle of truth. Myths and fables, as well as true stories about other people and other animals, often are repositories of truths of a very important sort. What I am denying is that what can be learned about people or the human condition from observing animals is any more warranted scientifically by the evidence adduced than is the "sour grapes" phenomenon by Aesop's fable. So far as elucidating human actions is concerned—as opposed to human bodily functions—animal studies are more like literary works than they are like scientific experiments. The author may have "gotten it right" for people as well as animals, but there is nothing in his presentation itself that assures us that he has.

As an illustration of this aspect of ethological discovery, we may consider an example from the work of Lorenz.[16] One of the things that he found in studying pair-formation in ravens is that it is the sex of a newly introduced prospective sexual partner that determines whether an individual that has been raised in isolation will act like a male or a female: regardless of its own biological sex, the isolated bird will adopt the courting behavior appropriate to the sex opposite to that of the introduced bird. Lorenz's finding, though intriguing, offers no basis for inference concerning any possible human situation. Any suggestiveness is merely implicit, that being part of the rhetorical effect of the example, though the result could very well express something that is also true of human beings. The suggestion is not altogether dissimilar to one that is found in D. H. Lawrence's story, "The Fox," in which the role that a young woman has assumed in a homosexual relationship is seen abruptly to change from that of a male to that of a female when a young man enters the scene. The writer and the ethologist are both in the position of being able to point out, but not to prove, something significant about the way the human creature behaves.

Biologically inspired claims concerning human social behavior, though lacking empirical validation, do nevertheless draw upon knowledge of what has been scientifically demonstrated. They are in that respect not unlike science fiction, wherein plausibility is conferred upon speculative propositions through showing them to be consistent with, but not entailed by, an established body of scientific theory. Unlike most science fiction, however, ethologically inspired propositions concerning humans commonly have a moral or political thrust, one that is all the more effective because of the illusion that these propositions have scientific backing. So far as biologists pretend to have shown that human behavior and human societies are constrained by biology to fall within limits that make it impossible, except through organic evolution, for human social patterns and social arrangements ever to be very different from what they are now, the only way to interpret the project of attempting to use biology to understand human social phenomena is as a way of rationalizing or defending the status quo. When a science is alleged to set forth for us what we can and cannot do because of our inherent nature, its pronouncements amount to either sound counsel or ideological preaching, depending on how well grounded in evidence the advice is. In the case of biology and human social behavior, it is both a scientific and a philosophical error to believe that the evidence warrants any conclusions either of a normative or descriptive kind concerning biological determination of either the form or the content of human social life. It is not by studying animals that we shall gain understanding of why people do as they do.

NOTES

1. Edward O. Wilson, *Sociobiology: The New Synthesis*. Cambridge: Harvard University Press, 1975, p. 4.
2. *Ibid.*, p. 551.
3. "Human Decency Is Animal," *New York Times Magazine*, Oct. 12, 1975, p. 48.
4. *Sociobiology*, pp. 564–65.
5. *Ibid.*, pp. 550–51.
6. *On Human Nature*. Cambridge: Harvard University Press, 1978, p. 152.
7. *Ibid.*, p. 153.
8. *Sociobiology*, p. 79.
9. *Ibid.*, p. 553.
10. See, e.g., N. Tinbergen, "On War and Peace in Animals and Men." *In*, Heinz Friedrich (ed.), *Man and Animal*. London: Paladin, 1972, pp. 118–42.
11. J. D. Carthy and F. J. Ebling (eds.), Institute of Biology Symposia No. 13: *The Natural History of Aggression*. New York: Academic Press, 1964, p. 5.
12. Konrad Lorenz, *On Aggression*. New York: Bantam Books, 1967, p. 229.

13. Robert Ardrey, *African Genesis*. London: Collins, 1961, p. 316.
14. *Sociobiology*, p. 564.
15. *Ibid.*, p. 225.
16. K. Lorenz, "Pair-Formation in Ravens." *In*, Friedrich (ed.), *Man and Animal*, pp. 17–36.

SOCIOBIOLOGY AND
BIOLOGICAL REDUCTIONISM

Sociobiology is a research strategy that seeks to explain human social life by means of the theoretical principles of Darwinian and neo-Darwinian evolutionary biology. Its aim is to reduce puzzles pertaining to the level of sociocultural phenomena to puzzles that can be solved on the biological level of phenomena. Biologists find sociobiology plausible and attractive because of its forthright commitment to the general epistemological principles of science. In this respect, cultural materialism and sociobiology are natural allies. In every other respect however, the two strategies are far apart. Cultural materialists of course accept neo-Darwinian principles when applied to the explanation of the social life of infrahuman species, but we insist that the same principles are capable of explaining only an insignificant proportion of human sociocultural differences and similarities. None of these puzzles can be solved effectively by means of the reductionist principles of sociobiology.

Basic Theoretical Principles of Sociobiology

In the neo-Darwinian evolutionary synthesis, the social behavior of different species of animals evolves as an outcome of differential reproductive success among individuals. Since the instructions for reproductive success are carried in the genes, one can view the entire course of biological evolution, including the evolution of patterns of animal social life, as the outcome of the preservation and propagation of the chemical organization of DNA.

Even among the simplest forms of organisms, however, behavior is not exclusively genetically determined. The actual observed response repertories of individual organisms result from the interaction of genetic instructions on the one hand, and the environment in which each organism is situated, on the other. Thus, each organism has a behavioral *genotype*—the ensemble of its hereditary instructions affecting behavior; and each organism has a behavioral *phenotype,* the genetically orchestrated product of its behavioral experiences in particular living sites. This is as true of human beings as of any other species as far as it goes.

Before the development of sociobiology, social behavior had presented a special challenge to classical Darwinian selection theory, because social life frequently involves reproductive costs that result in the lowering of individual "fitness" (i.e., the number of adult offspring in the next generation attributable to an individual's reproductive behavior). The sterile worker castes of insect societies are the extreme instance of such "altruism." According to classical evolutionary theory every organism is a means and a consequence of reproductive competition among individuals, and there should be no example of behavior benefiting another individual at reproductive cost to the performer (Eberhard, 1975). The puzzle of "altruism" was solved for infrahuman species with W. D. Hamilton's (1964) development of the concept of "inclusive fitness." Inclusive fitness explains genetically costly social acts in terms of the joint effect of such acts upon the fitness of individuals and their genetically related social partners. The closer the genetic resemblance between the altruistic individual and the benefited social partners, the greater the probability that altruistic behavior results in the preservation of the altruistic individual's genotype. Thus the development of the concept of inclusive fitness made it possible to explain the evolution of all genetically controlled variations in infrahuman animal behavior in conformity with the principle of natural selection.

It is understandable, therefore, why sociobiologists find the temptation to apply the same principle to the explanation of human social behavior well-nigh irresistible. Natural selection, however, has repeatedly been shown to be a principle under whose auspices it is impossible to develop parsimonious and powerful theories about variations in human social life (cf. Harris, 1968: 80 ff). The extension of natural selection to altruistic behavior in infrahuman social species in no way alters or diminishes the objections raised against other forms of biological reductionism, such as racism and instinctualism.

The Emergence of Culture

The weakness of human sociobiology and all other varieties of biological reductionism arises initially from the fact that genotypes never account for all the variations in behavioral phenotypes. Even in extremely simple organisms, adult behavior repertories vary in conformity with each individual's learning history. Hence the behavior repertory of conspecific social groups necessarily includes learned responses, and the prominence of such responses necessarily increases with the complexity of the neural circuitry of the species in question. These learned responses play an important role in the evolution of infrahuman social life, since they constitute a strategic part of each organism's behavioral phenotype. Selection acts on the behavioral phenotype, increasing the fitness of organisms with advantageous innovations in their response repertories. For example, a caterpillar accidentally captured by a wasp with an instinct for eating flies might be the first step toward the evolution of wasps equipped with an instinct for eating caterpillars. In classical evolutionary theory, therefore, behavior and genes form a positive feedback loop: innovative behavior is preserved and propagated by its contribution to the organism's fitness and is thus converted from an accidental byproduct of learning to a highly determined expression of the genotype.

Given sufficiently advanced forms of learning circuits, socially assisted learning is another and radically different process by which learned responses that have been found useful by one organism can be preserved and propagated within a social group. The social response repertories acquired by means of socially assisted learning constitute a group's tradition or culture. In principle, cultural repertories can change and evolve entirely independently of feedbacks involving natural selection—that is, entirely independently of the reproductive success of the individuals responsible for innovating and propagating them. For example, Thomas Alva Edison's invention of the phonograph would have spread around the world even if Edison and all his close kinspeople had been entirely childless.

Infra-Human Culture

There is nothing hypothetical or mysterious about culture. It did not come into existence through some sudden abrupt reorganization of the human mind; rather it emerged as a byproduct of the evolution of complex neural circuitry, and it exists in rudimentary form among many vertebrate species. But as in all evolutionary analysis, a balance must be struck between the continuities and discontinuities of emergent forms and processes. Sociobiologists underestimate by several orders of magnitude the extent to which human cultures represent an emergent novelty. They must therefore accept responsibility for disseminating a biased picture of the evolutionary processes affecting human societies. For human culture has doubtless fulfilled the theoretical potential of cultural evolution to a point that is absolutely unique among all organisms.

Examples of infrahuman culture are the dialects and call variations of birds and mammals; bird and vertebrate flyways, trails, display grounds, and nesting sites; and primate tool-using, feeding, and line-of-march specialties. The best-studied cases involve controlled colonies of Japanese macaques who responded to novel ways of provisioning by inventing novel ways of food processing. One troop developed the tradition of washing sweet potatoes in sea water and then went on to washing all its food in a similar manner. When wheat was thrown on the beach, members of the same troop discovered how to separate it from the sand by casting fistfuls of the mixture into the water and letting the grit sink to the bottom while scooping up the edible grains which floated on the surface. No genetic changes were required for these behavioral innovations to spread from individual to individual (Itani & Nishimura, 1973).

Under normal conditions, however, cultural innovations among infrahuman species are almost always brought back into the genetic feedback loop. If innovations persist at all, it is because they are valuable, and if they are valuable, they will affect reproductive success. For example, Wilson (1975:17) mentions an island-dwelling species of lizard (*Uta palermi*) which alone in its desert-adapted genus forages in the island's intertidal zones. The origin of this adaptation may have been purely behavioral, as in the case of the Japanese macaque innovations. But the lizards were soon genetically selected for their ability to forage at the seashore, and this innovative behavior became part of a genetically controlled species-specific behavioral repertory. In the Galapagos there are species of iguanas that swim and dive for their food, presumably as a consequence of a similar series of behavior-gene feedbacks.

The same kind of feedback probably accounts for many of the species-specific attributes of *Homo sapiens.* Among the primordial hominids of the Pliocene and Pleistocene periods, cultural innovations in tool use would undoubtedly have amplified trends leading toward the greater precision and power of the human thumb, and this in turn would have placed a selective premium on the neural circuitry needed for the intelligent use of hand-held tools. There is also little doubt that the development of *Homo sapiens'* unique language facility proceeded in a similar fashion, with selection favoring individuals able to transmit, receive, and store ever-more-complex messages. In the human case, however, these selection processes had a paradoxical outcome. In effect, by enhancing the capacity and efficiency of human learning functions, natural selection itself greatly reduced the significance of genetic feedback for the preservation and propagation of behavioral innovations. By progressively severing hominid cultural repertories from genetic coding, natural selection conferred an enormous adaptive advantage on *Homo sapiens*—namely, the advantage of being able to acquire and modify a vast range of useful behavior far more rapidly than is possible when genes maintain or regain control over each behavioral innovation.

The Evidence for Gene-Free Culture

How do we know that *Homo sapiens* has been selected for the capacity to acquire and modify cultural repertories independently of genetic feedback? The evidence for this viewpoint consists of the uniquely large amount of variation in the social response repertories of different human populations. Even the simplest of human societies exhibits tens of thousands of patterned responses not found in other human groups. George Peter Murdock's *World Ethnographic Atlas* contains forty-six columns of variable cultural traits. Over a thousand variable components per society can be identified by using the alternative codes listed under these columns. No two societies in the sample of 1,179 have the same combination of components. By adding more categories and by making finer distinctions, additional thousands of distinctive traits can be identified in the infrastructural sectors alone. In fact, lists employed by anthropologists interested in studying the phenomenon of diffusion include as many as six thousand traits (cf. Kroeber & Driver, 1932). The amphibious truck company of which I was a member during World War II had a supply manual with over one million items listed in its pages. A complete inventory of the material culture of U.S. society would certainly exceed a trillion items.

How do we know that these items are not part of a behavior-gene feedback loop? Because they can be acquired or wiped out within the space of a single generation, without any reproductive episodes taking place. For example, human infants reared apart from their parents in another breeding population invariably acquire the cultural repertory of the people among whom they are reared. Children of English-speaking American whites reared by Chinese parents grow up speaking perfect Chinese. They handle their chopsticks with flawless precision and experience no sudden inexplicable urge to eat McDonald's hamburgers. Children of Chinese parents reared in white U.S. households speak the standard English dialect of their foster parents, are inept at using chopsticks, and experience no uncontrollable yearning for bird's-nest soup or Peking duck.

Social groups and individuals drawn from a vast variety of populations have repeatedly demonstrated their ability to acquire every conceivable aspect of the world cultural inventory. American Indians brought up in Brazil incorporate complex African rhythms into their religious dances; American blacks who attend the proper conservatories readily acquire the distinctly non-African requisites for a career in classical European opera. Jews brought up in Germany acquire a preference for German cooking; Jews brought up in Yemen prefer Middle Eastern dishes. Under the influence of fundamentalist Christian missionaries, the sexually uninhibited peoples of Polynesia began to dress their women in dowdy Mother Hubbards and to follow rules of strict premarital chastity. Native Australians reared in Sydney show no inclination to hunt kangaroo, create circulating connubia, or mutilate their genitals; they do not experience uncontrollable urges to sing about witchetty grubs and the emu ancestors. The Mohawk of New York State came to specialize in construction trades and helped erect the steel frames of many skyscrapers. Walking across narrow beams eighty stories above street level, they were not troubled by an urge to build wigwams rather than office buildings. The rapid diffusion of such traits as sewing machines, power saws, transistor radios, and thousands of other industrial products points to the same conclusion. Acculturation and diffusion between every continent and every major race and micro breeding population prove beyond dispute that the overwhelming bulk of the response repertory of any human population can be acquired by any other human population through learning processes and without the slightest exchange or mutation of genes.

Among human beings, in other words, cultural life is not some sort of peripheral oddity. Each instance of a genuine cultural performance by a macaque or a chimpanzee is worth a journal article. But all the journals in all the libraries of the world would not suffice to render a running account of human cultural activities. Cultural evolution is thus responsible for creating an amount of intraspecific behavioral variation in the human species that does not exist in any other species. Moreover, this immense quantity of variation involves functional specialties whose analogues are associated with great phylogenetic distances in the evolution of other bioforms. The contrast between a paleotechnic foraging band and an industrial superpower is surely not inferior to the contrast between whole phyla—if not kingdoms—in the Linnaean taxonomy. It took billions of years for natural selection to create specialized adaptations for fishing, hunting, agriculture; for aquatic, terrestrial, and aerial locomotion; and for predatory and defensive weaponry, such as teeth, claws, and armor. Equivalent specialties were developed by cultural evolution in less than ten thousand years. The main focus of human sociobiology ought therefore to be the explanation of why other species have such minuscule and insignificant cultural repertories and why humans alone have such gigantic and important ones. But sociobiologists conceive their task to be something else— namely, the identification of the genetic components in human cultural traits. This represents a fundamental misdirection for human social science and a diversion of resources from the more urgent task of explaining the vast majority of cultural traits that do not have a definite genetic component.

The Scope of Sociobiological Theories

Popular representations of sociobiology have created a false impression of how sociobiologists relate human social behavior to its genetic substrate. Sociobiologists do not deny that most human social responses are socially learned and therefore not directly under genetic control. Wilson (1977:133) has made this point without equivocation: "The evidence is strong that almost but probably not quite all differences among cultures are based on learning and socialization rather than on genes." Richard Alexander (1976:6) has made the same pronouncement: "I hypothesize that the vast bulk of cultural variations among peoples alive today will eventually be shown to have virtually nothing to do with their genetic differences." Thus few if any sociobiologists are interested in linking variations in human social behavior to the variable frequencies with which genes occur in different human populations. The principles by which sociobiologists propose to implicate natural selection and reproductive success in the explanation of sociocultural phenomena are less provocative but also a good deal less precise and far less interesting.

Two such principles provide the auspices for most sociobiological theories concerned with human social life. The first of these seeks to explain the recurrence of certain universal or near universal traits as the consequence of a genetically programmed human nature; the second seeks to explain cultural variations as a consequence of a genetically programmed "scale" of alternatives which are allegedly turned on and off by environmental "switches."

Regarding the first principle, under the best of circumstances the search for human nature can yield only an understanding of the similarities in sociocultural systems and not of the vast repertory of differences. I shall demonstrate, moreover, that sociobiology is strategically biased in favor of exaggerating the number of pan-human traits that are part of human nature. This bias acts as a definite barrier to an adequate exploration of the causes of widely recurrent traits.

As for the second principle, I shall demonstrate that the sociobiological commitment to "behavior scaling"—the genetically predetermined traits turned on and off by environmental switches—introduces a superfluous set of variables into the solution of sociocultural puzzles. These puzzles are solved as soon as the environmental switches are adequately described in terms of the demo-techno-econo-environmental conjunction of the modes of production and reproduction. This aspect of sociobiology therefore emerges as simply a prolegomenon to cultural materialism.

Human Nature

Much confusion arises from the fact that sociobiologists present the concept of a genetically controlled "biogram," or human nature, as if there is a significant body of informed opinion asserting that human beings are not genetically programmed to be predisposed toward certain behavioral specialties. In principle there can be no disagreement that *Homo sapiens* has a nature. One does not have to be a sociobiologist to hold such a view. As every science fiction fan knows, a culture-bearing species whose physiology was based on silicon instead of carbon and that had three sexes instead of two, weighed a thousand pounds per specimen, and preferred to eat sand rather than meat would acquire certain habits unlikely to be encountered in any *Homo sapiens* society. The theoretical principles of cultural materialism hinge on the existence of certain genetically defined pan-human psychobiological drives that mediate between infrastructure and nature and that tend to make the selection of certain patterns of behavior more probable than others. Nothing I have said about the gene-free status of most cultural variations is opposed to the view that there is a human nature shared by all human beings. Hence the disagreement about the human biogram is entirely a matter of substance rather than of principle—that is, precise identification of the content of the biogram.

The disagreement between sociobiologists and cultural materialists on the issue of human nature is a matter of the contraction versus the expansion of the postulated substance of human nature. Cultural materialists pursue a strategy that seeks to reduce the list of hypothetical drives, instincts, and genetically determined response alternatives to the smallest possible number of items compatible with the construction of an effective corpus of sociocultural theory. Sociobiologists, on the other hand, show far less restraint and actively seek to expand the list of genetically determined traits whenever a plausible opportunity to do so presents itself. From the cultural materialist perspective, the proliferation of hypothetical genes for human behavioral specialties is empirically as well as strategically unsound, as I shall show in the next section.

Hypothetical Genes

According to Wilson (1977:132), we share a number of genetically controlled behavior traits with the Old World primates, while other traits are uniquely human. Among the more general primate traits Wilson mentions the following: (1) "size of intimate social groups on the order of 10-100"; (2) polygyny; (3) "a long period of socialization in the young"; (4) "[a] shift in focus from mother to age- and sex peer groups"; and (5) "social play with emphasis on role practice, mock aggression, and exploration." Among the features that are restricted to hominids, Wilson mentions: (6) facial expressions; (7) elaborate kinship rules; (8) incest avoidance; (9) "semantic symbol language that develops in the young through a relatively strict time table"; (10) close sexual bonding; (11) parent/offspring bonding; (12) male bonding; (13) territoriality. Wilson claims that "to socialize a human being out of such species-specific traits would be very difficult if not impossible, and almost certainly destructive to mental development" (ibid.).

But it is clear that most of those traits are not, on the ethnographic record, actually universal; and that socializing people out of them is no more difficult than socializing human beings out of thousands of other cultural traits.

1. Human relationships are at their most intimate in domestic groups. But the size of such groups varies over a far larger span than Wilson acknowledges. Needless to say, hundreds of millions of people in the world today live in domestic groups smaller than ten persons. While domestic groups larger than a hundred are less common, they do exist. For example, elite Chinese family compounds containing 700 persons are reported for the Sung dynasty (cited in Myron Cohen, 1975:227). But intimate relationships are not necessarily limited to co-residential domestic groups. Relationships within such nonlocalized kinship groups as lineages and kindreds may also satisfy definitions of "intimacy." Such groups commonly contain several hundred persons. In Brazil studies of elite families have demonstrated that the "parentella" or bilateral kindred may contain as many as 500 living relatives who keep track of one another's birthdays, attend marriages and funerals together, and help one another in business and professional life (Wagley, 1963:199). The size of such human groups as families, lineages, kindreds, villages, and "communities" varies in conformity with known infrastructural conditions and nowhere exhibits significant uniformities that render the existence of even minimal genetic restraint plausible.

2. True, polygyny is a common form of human mating, but the human species is not polygynistic. Human sexual behavior is etically so diverse as to defy any species-specific characterization. The heterosexual range runs from promiscuity through monogamy, with each type practiced by tens of millions of people. Human females in general may not have plural mates as often as males, but there are millions of women who entertain a plurality of sexual partners as often as, if not more often than, the most active men in other societies. This is particularly evident in the de facto forms of polyandry or rapid mate changes prevalent among the matrifocal households of the Caribbean and Northeast Brazil (Rodman, 1971; Lewis, 1966). Moreover, polyandry is an emic as well as an etic commonplace in Southwest India and Tibet. The idea that males naturally desire a plurality of sexual experiences while women are

satisfied by one mate at a time is entirely a product of the political-economic domination males have exerted over women as part of the culturally created, warfare-related male supremacy complex. Sexually adventurous women are severely punished in male-dominated cultures. Wherever women have enjoyed independent wealth and power, however, they have sought to fulfill themselves sexually with multiple mates with no less vigor than males in comparable situations. I cannot imagine a weaker instance of genetic programming than the polygyny of *Homo sapiens.* Sexuality is something people can be socialized out of only at great cost. But people can be socialized into and out of promiscuity, polygyny, polyandry, and monogamy with conspicuous ease, once the appropriate infrastructural conditions are present.

3. Human socialization does take a long time, and this indeed is part of *Homo sapiens'* primate heritage. But this feature of the biogram is closely related to the development of elaborate socially learned response patterns that constitute "traditions" or cultures. The trait of prolonged socialization merely points up the fact that human infants have a lot of traditions to learn and that *Homo sapiens'* most characteristic genetic heritage is the expanded capacity for cultural behavior. However, it does not point to any particular content for pan-human social life other than the existence of an extended training period for infants and children. What infants and children are trained to do is another story.

4. The alleged genetically controlled shift in focus from mother to age and sex peer groups is much more likely to be one of the things that *Homo sapiens* needs to be trained to do and that people can be socialized out of doing with little difficulty. Indeed, there is a very extensive anthropological and psychological literature suggesting that once strong dependency relationships are established between children and parents, costly training techniques are necessary to pry children loose and send them out into the world on their own. *Homo sapiens* is the only primate species that needs puberty rituals to shock and cajole the junior generation into accepting adult responsibilities (Harrington & Whiting, 1972). Moreover, it is a caricature of human socialization processes to represent them primarily in terms of mother-child relationships, since human fathers often play as important a role in socialization as mothers

do. And both human parents frequently continue to dominate the activities of their offspring well into adulthood. If Wilson's notion of primate genetic control over human role socialization were correct, the system of financing higher education through twenty years or more of heavy parental investment would never have been invented.

5. *Homo sapiens* does share with other primates a genetically programmed tendency toward exploratory social play in connection with role practice. However, this feature is redundant with respect to item 3—the prolonged nature of human socialization—and merely points once again to the importance of socially acquired response repertories rather than to the importance of genetic control over definite response categories. On the other hand, Wilson's specification of "mock aggression" as a genetically controlled feature of childhood social play lacks credibility. Children explore the social roles and behavior patterns their cultures encourage them to explore. As formalized in sports and games, the balance between aggressive and nonaggressive childhood behavior patterns varies widely from culture to culture. Competitive team sports for males, for example, are correlated with training for warfare (Sipes, 1973). In societies lacking warfare patterns, such as the Semai of Malaysia, children's play is quite free of conspicuous aggressive displays (Dentan, 1968; Montagu, 1976:98–103). Obviously human beings have the genetically controlled capacity to act aggressively, but the conditions under which aggressive behavior actually manifests itself is not defined by a narrow set of genetic instructions.

6. Facial expressions probably constitute one of the best cases for definite genetically controlled response patterns. There is a universal tendency for *Homo sapiens* to laugh and smile in order to communicate pleasure, to frown and stare when communicating anger, and to grimace and cry when communicating pain and sorrow. Even in this case, however, the genetic programming cannot be very strong, since many cultures override the species-specific meanings and use the same facial expression to denote something quite different. All over the world people are socialized to hide their feelings and to laugh when they are sad, look forlorn when they are happy, and smile when they are angry. It is customary in many Amazonian Indian societies to weep profusely in honor of arriving

guests. Elsewhere, as in the Middle East and India, the rich pay professional mourners to weep at funerals. Wherever it is important not to show true feelings—as in the game of poker or during job interviews—people learn to master their facial muscles just as readily as they learn bladder and sphincter control and with much less threat to their mental and physical well-being. In most societies it is dangerous to rely on facial expressions in order to predict what people are getting ready to do.

7. True enough, complex kinship rules are not found in any other species, but the variety of kinship terminologies, family organization, and prescribed behavior for kinspeople is far too great to be accounted for by genetic controls. These rules are "elaborate" precisely because they are not under genetic control. What kind of gene would it be that could lead some societies to distinguish cross- from parallel cousins or mother's brother from father's brother? Or that would give the members of Australian aboriginal groups in the eight-section zones an uncontrollable urge to marry their mother's mother's brother's daughter's daughter? Moreover, despite their universality in one form or another, kinship rules need not be regarded as permanent features of the universal pattern. The entire history of state society is one convergent thrust toward the replacement of kin-organized groups by those based on the division of labor, class, and other achieved statuses.

8. That the widespread prohibition on mother-son, father-daughter, and brother-sister mating is genetically programmed I find dubious. Brother-sister mating was practiced by the Egyptian pharaohs, the ruling elites of Hawaii, the Inca, and the early Chinese emperors as a routine means of consolidating power at the apex of the social pyramid. On the other hand, there are excellent infrastructural reasons why brother-sister marriage was prohibited virtually everywhere else, since all alliances and exchanges between commoner domestic groups are based on brothers not taking their sisters for themselves. One expects father-daughter marriages to be rare for the same reason. However, sexual relations between father and daughter take place on a rather frequent basis, sometimes in overt ritual contexts. If incest avoidance is part of the human biogram, why are there, conservatively estimated, several hundred thousand cases of father-daughter incest in the United States each year (Armstrong, 1978:9)?

The rarest form of nuclear-family incest is the mother-son relationship. The problem here is to disentangle an instinctual repugnance from the generally subordinate status of women and the rules against adultery. Does fear of father and husband or fear of incest per se keep sons and mothers apart? If psychoanalytic evidence is admissible, then it is clear that in childhood and adolescence sons do have a rather powerful sexual interest in their mothers. That interest, however, has to be blocked if sons are to develop in the image of their aggressive masculine fathers. Because of the prevalence of male supremacy women are usually awarded to men for achievement and not as a birthright. Hence virtually all societies currently have an interest in preventing mother-son incest as part of the process of training males into their masculine role. Very probably, as the premium on training males for aggressive masculine roles diminishes, less effort will, be expended on discouraging sons from fulfilling their Oedipal fantasies, and the incidence of mother-son incest may increase. As long as the infrastructural and structural conditions sustaining the nuclear-family incest taboos remain in force, the conclusion that the taboos are instinctual remains unconvincing (cf. Y. Cohen, 1978). Only if the taboos persist after the sustaining infrastructural and structural conditions disappear will there be convincing evidence of their genetic base. Moreover, no argument using the antiquity of the incest taboos as evidence of their instinctual status is admissible. The hunter-collector mode of production lasted a million years or more, yet it was swept away in a mere handful of generations when changed infrastructural conditions altered the balance of costs and benefits.

9. The human capacity to communicate by means of a "semantic symbol language" does involve a genetically programmed predisposition to acquire such a language, and it is definitely known that no other species on earth shares the same predisposition. But the inclusion of this one momentous instance of genetic specialization in a list of dubious or untrue features of the human biogram reveals the weakness of the sociobiological strategy. For the behavioral implication of the unique language facility of human beings is that *Homo sapiens* has a unique, genetically based capacity to override genetic determinism by acquiring, storing, and transmitting gene-free repertories of social responses. I shall return to this point in a moment.

10. Close sexual bonding? How can one assert this as a species-specific feature when under appropriate infrastructural and structural conditions, the sexual bond extends no further than rape, prostitution, and slave-forms of concubinage?

11. Parent/offspring bonding? Sex ratio studies indicate that infanticide, especially female infanticide, and neglect of young girls were routine means of regulating human population growth throughout history as well as prehistory.

12. Male bonding? To be sure, cross-culturally, solidary male groups outnumber solidary female groups. But that merely reflects once again the cultural tendency of males to keep females subordinate by preventing them from forming aggressively solidary coalitions. To suggest that women cannot form strong solidary bonds among themselves because they are "catty" and preoccupied with what men think of them (cf. Tiger, 1970) has little merit. There is enough ethnographic evidence to indicate that women can and do form effective solidary political groups, as in the Bundu of Sierra Leone (Hoffer, 1975) as well as in the more recent burgeoning of feminist associations. Female sodalities will undoubtedly become more common as women achieve political and economic parity with men.

13. Territoriality? Recent studies of hunter-gatherers support the theory that the primordial units of human social life were open camp groups whose membership fluctuated from season to season and whose territories were not sharply defined. Strong territorial interests probably became widespread only after the development of sedentary village modes of production and the rise in reproductive pressure. Moreover, the ownership of territory by groups or individuals is certainly not something that it is difficult to socialize people out of, nor is the lack of territorial interests "destructive of mental development." (Indeed, one might argue more cogently that the territorial interests of modern states are destructive not only of mental development but of physical existence as well.)

Sociobiological Misrepresentation of Human Nature

Sociobiologists argue quite persuasively that *"Homo sapiens* is distinct from other primate species in ways that can be explained only as a result of a unique human genotype" (Wilson, 1977:132). Yet when it comes down to the details of characterizing that genotype, sociobiologists overlook or minimize the genetic trait that by their own criteria ought to be emphasized above all others. That trait is language. Only human language has semantic universality—the ability to communicate about infinite classes of events regardless of when and where they occur. This trait and the neural circuitry that makes it possible account for the astonishing fact—agreed to by most sociobiologists—that the enormous within-species variations in human social response repertories are not genetically controlled.

The attempt by sociobiologists to add what are at best dubious and hypothetical genes to the human behavioral genotype leads to the misrepresentation of human nature based on an erroneous construal of the course of hominid evolution. According to the latest paleontological evidence, the phyletic lines of the ancestral hominids and their closest pongid relatives have been separated for at least 5 million years. During all that time, natural selection favored a behavioral genotype in which the programming acquired through learning progressively dominated the programming acquired through genetic change. Every discussion of human nature must begin and end with this aspect of the human biogram, for its importance overrides every other conceivable species-specific trait of *Homo sapiens*. Indeed, the emergence of semantic universality constitutes an evolutionary novelty whose signficance is at least as great as the appearance of the first strands of DNA. As I have said, culture appears in rudimentary form among the lower organisms, but it remains rudimentary; it does not expand and accumulate; it does not evolve. Hominid culture, however, has evolved at an ever-increasing rate, filling the earth with human societies, artifacts, and countless divergent, convergent, and parallel instances of behavioral novelties. Even if the dubious hypothetical genetic predispositions of sociobiological human nature actually do exist, knowledge of their existence can lead only to an understanding of the outer "envelope" (to use a metaphor proposed by Wilson—Wilson & Harris, 1978) within which cultural evolution has thus far been constrained. It could not lead to an understanding of the differences and similarities within sociocultural evolution. The poverty of this strategy can best be grasped by imagining what evolutionary biology would be like if it confined itself only to the similarities of life forms and not to their differences. Darwin's contribution would then amount to nothing more than the assertion that all species are constrained by their common carbon chemistry and by the laws of thermodynamics. The question of why whales are different from elephants or why birds are different from reptiles could not be asked. Or at least it wouldn't pay to ask it, since the answer would be: "They are not really different: they are all constrained within the envelope of carbon chemistry and thermodynamics."

The Principle of Behavior Scaling

In order to overcome the apparent irrelevance of genetic controls for the explanation of sociocultural differences, sociobiologists turn to the concept of "behavioral scale" (Wilson, 1975:20–21). "Scaling . . . refers to those cases in which the genetic code programs not for invariant phenotypic response, but for variant, but predictable responses to varying environmental conditions . . ." (Dickeman, 1979:1). The most familiar examples of behavioral scaling involve responses dependent on changes in population density. For example, aggressive encounters among adult hippopotami are rare where populations are low to moderate, but at high densities, males begin to fight viciously, sometimes to the death. Normally, snowy owls do not engage in territorial defense, but under crowded conditions they defend their territories with characteristic displays. Availability of food also triggers different parts of the behavior scale. Well-fed honeybees let intruding workers from neighboring hives penetrate the nest and take away supplies without opposition. But when the same colonies have been without food for several days, they attack every intruder. "Thus the entire scale, not isolated points on it, is the genetically based trait that has been fixed by natural selection" (Wilson, 1975:20).

Applied to human social response repertoires, the principle of behavior scaling is meant to provide an explanation for why some human populations are polygynous while others are polyandrous, or why some human populations are cannibals while others are vegetarians, and so forth. These alternative behaviors constitute a genetically programmed "range of possible behavioral responses" evoked by particular ecological contexts.

The Scope of Behavioral Scaling

The premise that cultural innovations are preprogrammed along a species-specific scale suffers from the strategic liabilities associated with the notion of human nature. Since selection in the main has acted against genetically imposed limitations on human cultural repertories, the principle of scaling leads to an evolutionarily unsound strategy in the human case. Once again the problem is that the diversity of human responses is far too great to be accounted for by a genetic program. Unlike hippopotami, human beings do not merely change from being placid to being aggressive and back again under the variable stress of environment. Instead, humankind has evolved an enormous series of institutions such as lineages, polygyny, polyandry, redistribution, slavery, sacrifice of prisoners of war, infanticide, vegetarianism, and all the other infrastructural, structural, and superstructural traits. Clearly the concepts of behavior scaling cannot be applied to this entire series of innovations. To do so would in effect reduce all cultural evolution to a predetermined set of genetic instructions, a conclusion at odds with the agreed-upon fact that most human behavior is not under direct genetic control. For example, consider the evolutionary series: bands, village, chiefdom, state. Surely no one would want to suggest that these alternative forms of political economy are merely preprogrammed points on a genetically controlled behavior scale. That would be the equivalent of saying selection already favored the state even before there were chiefdoms or villages and when only bands existed. Behavior scaling, in other words, still leaves the explanation of the origin of most of the diversity of human social life beyond the pale of sociobiological theory, and hence does not alter the relative lack of scope and power built into the sociobiological approach to human social life.

The Superfluous Gene

To an important but limited extent, the principle of behavior scaling resembles the cultural materialist principle of infrastructural determinism. What the two principles have in common is the specification of an ecological context and the assumption that variations in social response repertories are in some sense "adaptive" to that context. In the cultural materialist strategy the ecological conditions embodied in the infrastructure raise or lower the bio-psychological costs and benefits of innovative responses, none of which is genetically preprogrammed to occur under the given conditions. Certain innovative behaviors rather than others are retained and propagated not because they maximize reproductive success but because they maximize bio-psychological benefits and minimize bio-psychological costs. In contrast, the fundamental premise of the principle of scaling is that the innovative behavior is genetically preprogrammed to occur under the given conditions because under the given conditions it maximizes reproductive success.

I contend that the principle of behavior scaling leads at best to unparsimonious theories concerning the manner in which innovations come to take their place in the human behavioral repertory. The lack of parsimony results from the need to rely on hypothetical genes in order to fulfill the genetic cost-benefit optimizing or mini-max logic of the sociobiological strategy. The logic of the cultural materialist optimizing mini-max, cost-benefit analysis, on the other hand, is fulfilled in a more direct fashion and is to be preferred on that account. Indeed, from the cultural materialist perspective, the invocation of genetic factors in behavior scaling explanations constitutes a redundant and gratuitous addition to an adequate ecological and psycho-biological cost-benefit analysis.

The Case of Elite Female Infanticide

As an example of the lack of parsimony and the redundancy involved in the behavior scaling version of sociobiological theory, let us examine the explanation anthropologist Mildred Dickeman (1979) has proposed for the occurence of female infanticide among elite castes and classes of late medieval Europe, India, and China. To explain this important and empirically valid phenomenon, Dickeman relies on a sociobiological model developed by Richard Alexander (1974), which predicts that female preferential infanticide is more likely among women married to high-ranking men and less likely among women married to low-ranking men. The logic behind Alexander's model is this: when male infants can be reared with confidence, their fitness (i.e., number of offspring) will tend to exceed that of females, since men can have many more reproductive episodes than women. Hence in elite castes and classes, where males have an excellent chance of surviving because living conditions are good, the maximization of reproductive success of both male and female parents will be achieved by investing in sons rather than daughters. On the other hand, in the low-ranking castes and classes, where male survival is very risky, reproductive success will be maximized by investing in daughters, who are likely to have at least some reproductive episodes rather than none at all. To complete the model, elite men can be expected to marry beneath their station, while the lowly women can be expected to marry up if their parents can provide them with a dowry to compensate the groom's family.

The cultural materialist explanation for the occurrence of female infanticide in elite groups dispenses with the need to suppose that this pattern has been selected for genetically and that it is part of a genetic program that automatically manifests itself under conditions of extreme poverty and wealth. We begin with the fact that daughters were less valuable than sons for the Eurasian elites because men dominated the political, military, commercial, and agricultural sources of wealth and power. (Male politico-economic domination itself is also a product of cultural rather than genetic selection.) Sons have the opportunity to protect and enhance the elite family's patrimony and political-economic status. But daughters, who have access to significant sources of wealth and power only through men, are an absolute or relative liability. They can only be married off by paying dowry. Therefore, preferential female infanticide is practiced by the elite groups to avoid the expense of dowry and to consolidate the family's wealth and power. Among the subordinate ranks, female infanticide is not practiced as frequently as among the elites because peasant and artisan girls can readily pay their own way by working in the fields or in cottage industries.

The genesis of this system lies in the struggle to maintain and enhance differential political-economic power and wealth, not in the struggle to achieve reproductive success. The proof of this lies in the outcome of the hypogynous marriages (marrying down) of the elite males. Such marriages generally take the form of concubinage and do not bestow the right in inheritance upon the offspring, be they male or female. The elites, in other words, systematically decrease their inclusive fitness by failing to provide life-support systems for their own children. Indeed, I would argue that the entire complex of infanticide and hypergamy boils down to a systemic attempt to prevent the elites from having too much reproductive success in order to maintain the privileged position of a small number of wealthy and powerful families at the top of the social pyramid.

Who Needs Behavior Scaling?

Sociobiologists propose that human beings are preprogrammed to switch from infanticide to mother love; from cannibalism to vegetarianism; from polyandry to polygyny; from matrilineality to patrilineality; and from war to peace whenever the appropriate environmental conditions are present. Cultural materialists also maintain that these changes take place whenever certain infrastructural conditions are present. Since both cultural materialists and sociobiologists take the position that the enormous diversity represented in the alleged genetic scaling of human responses is at least genetically possible—within the "envelope"—the need for the scaling concept itself seems gratuitous. The focus in both strategies has to be on the question of what kinds of environmental or infrastructural conditions are powerful enough to change human behavior from war to peace, polygyny to polyandry, cannibalism to vegetarianism, and so forth. To the extent that sociobiologists sincerely pursue this issue, they will inevitably find themselves carrying out cost-benefit analyses that are subsumed by the infrastructural cost-benefit analyses of cultural materialism.

True, sociobiological models based on reproductive success and inclusive fitness can yield predictions about sociocultural differences that enjoy a degree of empirical validity, as in the above case of infanticide and female hypergamy. But the reason for this predictability is that most of the factors that might promote reproductive

success do so through the intermediation of bio-psychological benefits that enhance the economic, political, and sexual power and well-
being of individuals and groups of individuals. The exploitation of
lower-ranking women by higher-ranking men, for example, is the
kind of stuff out of which theories of reproductive success can easily
be spun. But exploitation confers much more immediate and tangi-
ble benefits than genetic immortality on those who can get away
with it. Because of the bias toward reproductive success, the principle of behavior scaling leads away from the most certain and powerful interest served by infrastructure toward the most remote and
hypothetical interests served by having genetic survivors. Thus sociobiology contributes to the obfuscation of the nature of human social life by its commitment to the explorations of the least probable
causal relationships at the expense of the most probable ones.
However, it is not the sociobiologists who are primarily to blame
for this situation.

Who Is To Blame for Sociobiology?

The decisive consideration concerning the appropriateness of genetic
models for human behavioral repertories must be whether or not
there are cultural, nongenetic theories that account better for the
observed phenomena. Sociobiological theories are welcome only in
the absence of plausible sociocultural theories, because the latter derive from principles capable of explaining both rapid and slow
changes and both similarities and differences, whereas the former
derive from a principle—natural selection—capable of explaining
only slow changes and few if any of the differences as well as the
similarities. Anthropologists who operate with synchronic, idealist,
structuralist, and eclectic research strategies incapable of producing
interpenetrating sets of theories about the divergent and convergent
trajectories of sociocultural evolution have only themselves to
blame if sociobiologists step in to what appears to be an intellectual
disaster area. Sociobiology has achieved instant popularity in part
because the better-known social science research strategies cannot
provide scientific causal solutions for the perennial puzzles surround-
ing phenomena such as warfare, sexism, stratification, and cultural

life-styles. Sociobiologists have been accused of being racists and sexists and have been verbally abused at scientific meetings by academics who are committed to obscurantist and explicitly antiscientific strategies. This abuse at the hands of people who have made no contribution of their own to the explanation of sociocultural phenomena can only serve to strengthen the conviction of the sociobiologists that they are being martyred for their devotion to the scientific method and that the study of the most momentous issues in human life has been monopolized by those least competent to carry it out. The answer to sociobiology does not lie in still more abuse; it lies in the development of a corpus of coherent sociocultural theory that is more parsimonious and that has greater scope and applicability than the corpus of theories produced under the auspices of sociobiological principles. In the words of Imre Lakatos (1970:179): "Purely negative criticism does not kill a research program."

BIBLIOGRAPHY

Alexander, Richard 1974 "Evolution of Social Behavior." *Annual Review of Ecological Systems,* 5:325–83.
——— 1976 "Evolution, Human Behavior, and Determinism." *Philosophy of Science Association—Proceedings of the Biennial Meetings,* 2:3–21.
Armstrong, Louise 1978 *Kiss Daddy Goodnight.* New York: Hawthorn.
Cohen, Myron 1975 *House United, House Divided: The Chinese Family in Taiwan.* New York: Columbia University Press.
Cohen, Yehudi 1978 "The Disappearance of the Incest Taboo." *Human Nature,* 1(7):72–78.
Dentan, Robert 1968 *The Semai: A Non-Violent People of Malaya.* New York: Holt, Rinehart & Winston.
Dickeman, Mildred 1979 "Female Infanticide and the Reproductive Strategies of Stratified Human Societies: A Preliminary Model." *In,* Napoleon Chagnon and William Irons (eds.), *Evolutionary Biology and Human Social Behavior: An Anthropological Perspective.* North Scituate, Mass.: Duxbury (1979).
Eberhard, Mary Jane 1975 "The Evolution of Social Behavior by Kin Selection." *The Quarterly Review of Biology,* 50:1–33.

Hamilton, W. D. 1964 "The Genetical Evolution of Social Behavior." *Journal of Theoretical Biology*, 7:1–52.

Harrington, Charles, and J. Whiting 1972 "Socialization Process and Personality." *In,* Francis Hsu (ed.), *Psychological Anthropology.* Cambridge, Mass.: Schenkman, pp. 469–507.

Harris, Marvin 1968 *The Rise of Anthropological Theory.* New York: T. Y. Crowell.

Hoffer, Carol 1975 "Bundu: Political Implications of Female Solidarity in a Secret Society." *In,* Dana Raphael (ed.), *Being Female: Reproduction, Power, and Change.* The Hague: Mouton, pp. 155–64.

Itani, J., and A. Nishimura 1973 "The Study of Infrahuman Culture in Japan: A Review." *In,* E. W. Menzel (ed.), *Precultural Primate Behavior.* Basel: S. Karger, pp. 26–50.

Kroeber, Alfred, and H. Driver 1932 "Quantitative Expression of Cultural Relationships." *University of California Publications in Archaeology and Ethnology,* 29:252–423.

Lakatos, I. 1970 "Falsification and the Methodology of Scientific Research Programmes." *In,* I. Lakatos and A. Musgrave (eds.), *Criticism and the Growth of Knowledge.* Cambridge: Cambridge University Press, pp. 91–195.

Lewis, Oscar 1966 *La Vida.* New York: Random House.

Montagu, Ashley 1976 *The Nature of Human Aggression.* New York: Oxford University Press.

Murdock, George 1976 *World Ethnographic Atlas.* Pittsburgh: University of Pittsburgh Press.

Rodman, Hyman 1971 *Lower-Class Families: The Culture of Poverty in Negro Trinidad.* New York: Oxford University Press.

Sipes, Richard 1973 "War, Sports, and Aggression: An Empirical Test of Two Rival Theories." *American Anthropologist,* 75:64–86.

Tiger, Lionel 1970 *Men in Groups.* New York: Vintage.

Wagley, Charles 1963 *An Introduction to Brazil.* New York: Columbia University Press.

Wilson, E. O. 1975 *Sociobiology: The New Synthesis.* Cambridge: Harvard University Press.

——— 1977 "Biology and the Social Sciences." *Daedalus,* 106(4): 127–40.

Wilson, E. O., and Marvin Harris 1978 "The Envelope and the Twig." *The Sciences,* 18:10–15, 27.

A PROFFERING OF UNDERPINNINGS

On Human Nature completes, its author tells us, an unplanned trilogy, which, starting with insect societies, progressed through the sociobiology of vertebrates and is here concluded as a speculative essay on the application of sociobiology to the study of human affairs. Sociobiology is here defined as a "hybrid discipline that incorporates knowledge from ethology . . . , ecology . . . , and genetics in order to derive general principles concerning the biological properties of entire societies." Wilson's thesis is that without the underpinning provided by such principles the humanities and social sciences are doomed to remain ineffectual, unable to provide more than limited descriptions of superficial phenomena, with no real understanding of underlying causes. It is a thesis not calculated to endear its author to all his readers.

The first, and central, proposition is that our social behavior is to a significant extent genetically determined. "The accumulated evidence for a large hereditary component is more detailed and compelling than most persons, including even geneticists, realize. I will go further: it already is decisive." What does this statement mean? And why should we accept it? Let us consider the evidence first. Wilson points to some rather obvious ways in which human society is affected by human nature. It can hardly be doubted, for example, that our capacity for language is in some sense dependent

From *Science*, Vol. 204, pp. 735–37, May 18, 1979. Copyright 1979 by the American Association for the Advancement of Science. Reprinted by permission.

on our genetic makeup and that human culture has been profoundly affected by that capacity, or that most social organizations reflect, in one way or another, the length of an infant's dependence on its parents. Even the most determined opponent of sociobiology would presumably accept that human society reflects some biological facts such as these. If this were all Wilson had in mind, there would be little to argue about, and *On Human Nature* would be a very dull book.

But it is not. We are soon given more exciting fare. Wilson sees evidence of genetic determination in a wide, not to say haphazard, array of supposed facts. Some examples will give a flavor of the argument. First, there are characteristics we share with other higher primates. Thus "our intimate social groupings contain on the order of ten to one hundred adults, never just two, as in most birds and marmosets, or up to thousands, as in many kinds of fishes and insects." And the difference in physical size between men and women leads to the prediction, based on interpolation from other primates, that the "average number of females per successful male" should be greater than one but less than three. "The prediction," we are assured, "is close to reality; we know we are a mildly polygynous species." Where our social behavior differs from that of other primates, Wilson sees evidence of genetic determination in cultural universals. Traits such as bodily adornment, dancing, interpretation of dreams, law, medicine, personal names, and trade are "as diagnostic of mankind as are distinguishing characteristics of other animal species—as true to the human type, say, as wing tesselation is to a fritillary butterfly."

The reader may decide for himself or herself whether these arguments are convincing. What one misses from Wilson's account is a serious and sustained attempt to analyze in just what sense our behavior or social organization is genetically determined, and just how certain genetically determined characteristics have worked themselves out so as to produce, let us say, such cultural universals as law. Even more distressing is the shifting sense given to the notion of genetic determination. Wilson is cautious at times: our genes do not always dictate our lives too narrowly. "Rather than specify a single trait, human genes prescribe the *capacity* to develop a certain array of traits." And he knows full well that social organization varies enormously across time and place and that the precise form of a particular trait is often, perhaps usually, culturally determined. But although at the outset of his discussion he provides an unexceptionable definition of a genetically determined trait as one "that differs from other traits at least in part as a result of the presence of one or more distinctive genes," he soon slips into the habit of equating genetic determination with genetic constraint. "Human nature is stubborn, and cannot be forced without a cost." And, "It is inconceivable that human beings could be socialized into the radically different repertoires of other groups such as fishes, birds, antelopes, or rodents. . . . To adopt with serious intent, even in broad outline, the social system of a nonprimate species would be insanity in the literal sense. Personalities would quickly dissolve, relationships disintegrate, and reproduction cease."

This, let us remember, is a scientist writing. But what sort of scientific statements are these? And why should we believe them? It is not merely that Wilson offers no evidence; he does not pause to consider what sort of evidence might persuade us that our personalities would disintegrate and reproduction cease if we adopted, for example, the social organization of birds.

A similar tension between the rash and the cautious, the outrageous and the dull, the provocatively reactionary and the orthodoxly liberal, is evident throughout the book. Many of Wilson's particular arguments seem nicely calculated to ruffle left-wing sensibilities: warfare is a consequence of ethnocentrism; we are all innately aggressive, although women less so than men; while, still on the subject of sex differences, girls are naturally more sociable than boys. But, before he can be easily typed, Wilson adjures us to prize human diversity rather than discriminate against minorities, to regard homosexuality not only as biologically normal but also as biologically valuable.

The picture of Wilson as a crypto-fascist, painted with such abandon after the publication of *Sociobiology,* is without question a crude and mischievous caricature. In the last analysis, the major defect revealed by the present book is not a matter of political stance; much more serious is an all-pervading confusion as to the nature of the arguments being advanced and the conclusions that may legitimately be drawn from them. Wilson does not appear to see, for example, that pointing to the adaptive significance of a particular trait or the evolutionary advantage conferred by a particular pattern of behavior or social organization will not necessarily tell us anything else about that trait and certainly does not rule out, supplant, or allow one to choose between various other classes of explanation. To take an example almost at random: Wilson is concerned to press a biological explanation of the prevalence of incest taboos in preference to a sociological or cultural explanation. According to "the prevailing sociobiological explanation," such functions as preservation of the integrity of the family or facilitation of bridal bargaining, favored by anthropologists as explanations of incest taboos, are "by-products or at most secondary contributing factors." The sociobiolological explanation "identifies a deeper, more urgent cause, the heavily physiological penalty imposed by inbreeding." But biological explanations simply cannot preempt sociological explanations in this way. We could accept that the biological function of incest taboos was to prevent inbreeding, but this would not exclude the possibility that they served social functions also and not just as secondary, contributing factors. Nor would the identification of any such biological function allow one to choose between an infinity of possible proximate (social or psychological) causes. And finally, it would not even give any grounds for supposing that the behavior in question was in any significant sense genetically determined.

This failure to stress the limits—logical and conceptual rather than empirical limits—of sociobiological explanations is not confined to confusion between different levels of scientific discourse. Wilson equally tends to obscure the distinction between scientific explanation and judgment of value or political, social, or ethical belief. In a chapter on the sociobiology of religion, for example, the claim that "the highest forms of religious practice . . . confer biological advantage" is said to show that "religion itself is subject to the explanations of the natural sciences," to be treated as "a wholly material phenomenon." But even if we accepted that religion serves some biological function (and it is entirely characteristic that Wilson offers no evidence for such a proposition), what then follows? The answer is surely very little—certainly not that any of the claims made by or on behalf of religion are to be doubted. If the truth of a proposition is logically independent of one's motives for asserting it, it is even less dependent on the biological advantage its assertion may confer.

Evolutionary explanations, even if very much better founded than most of those on offer here, do not contradict other categories of explanation. Still less do they dictate ethical judgments. The gap between "is" and "ought" remains as wide today as it was when Hume pointed it out 200 years ago. Although Wilson pays lip service to this distinction between matters of fact and matters of value, we may be permitted to doubt whether his heart is in it. Had he taken it more seriously, he would never have asked us to believe "that a correct application of evolutionary theory . . . favors diversity in the gene pool as a cardinal value." Genetic diversity may be the stuff of which evolution is made, but evolution is not a cardinal value. And the particular implication drawn from this remark is distinctly weak: "If genius is to any extent hereditary, it winks on and off through the gene pool in a way that would be difficult to predict . . . For this reason alone, we are justified in considering the preservation of the entire gene pool as a contingent primary value." Even a lukewarm eugenist should not find it difficult to demolish that line of reasoning.

As a final example, and as if to provide the more strident critics of *Sociobiology*'s inferred political message with the ammunition they wanted, Wilson assures us that science has pronounced the sentence of death on Marxism. Marxism is "mortally threatened by the discoveries of human sociobiology," since it relies on "hidden premises about the deeper desires of human beings and the extent to which human behavior can be molded by social environments. These premises have never been tested. To the extent that they can be made explicit, they are inadequate or simply wrong." One wonders how we could know them to be wrong if they have never been tested. One also wonders what Wilson understands by Marxism. It seems most likely that he has equated it with the more foolish remarks of his radical American critics. The fact that people who call themselves Marxists have held certain views, for example on human genetics and the malleability of human behavior, does not imply that such views are a logical corollary of Marxism as a theory. It would not even follow had Marx and Engels themselves held such views—although in fact they did not. The social and political implications of biological facts are more complex and less determinate than is often thought. They deserve much more serious consideration than Wilson gives them—if only because of the increasingly widespread belief that the facts, theories, and speculations of human biology carry a message that will be welcome only to the most reactionary conservative. Wilson can hardly be blamed for this: he is, indeed, much less culpable than his radical critics, whose opposition makes it clear that they either share this belief themselves or else assume that the rest of the world does.

What can one say for *On Human Nature?* In all conscience, not much. Wilson is a distinguished scientist, but neither the distinction nor the science is much in evidence here. Where he offers us scientific hypotheses, he rarely provides the evidence to support them, and frequently does not stop to consider how one might set about collecting relevant evidence. Where he offers us his political and social views, he does so as though the respect due to science would lend them added authority. It does not.

PARIS:
MOSES AND POLYTHEISM

"Almost two thousand years, and no new god!"
Nietzsche, *The Antichrist*

Voltaire said that if God did not exist, man would have to invent Him. If we are to believe the French press, 1979 may be remembered as the year when two very different Parisian intellectuals applied for their respective patents on their own brand of deity.

With *Le testament de Dieu,* Bernard-Henri Lévy, thirty-one years old, ex-Maoist, ex-journalist, and self-proclaimed "New Philosopher," has become the latter-day prophet of a God who, though now deceased, was kind enough to leave behind His last will and testament, the Bible, as a bulwark against totalitarianism. With *Les idées à l'endroit* Alain de Benoist, ex-Catholic, ex-reactionary, and self-proclaimed "theoretical journalist," has presented a compendium of essays that attempts to lay the sociobiological foundations for a new paganism, a new aristocracy, and what is called the "New Right." "The debate between monotheism and polytheism," de Benoist writes, "is a truly essential discussion." But strangely enough, neither man actually *believes* in the deity or deities he proposes: they are merely convenient foils to help man muddle through the mess of the modern world. Nietzsche was

Review of *Barbarism with a Human Face* by Bernard-Henri Lévy, translated by George Holoch (Harper & Row); *Le testament de Dieu* by Bernard-Henri Lévy (Grasset); *Les idées à l'endroit* by Alain de Benoist (Hallier); *Vu de droite* by Alain de Benoist (Copernic).

From *The New York Review of Books,* January 24, 1980. Copyright © 1980 NYREV, Inc. Reprinted by permission of *The New York Review of Books.*

right after all. You can take your pick: the barren heights of Mount Sinai with Lévy, or the misty haunts of Celtic forests with de Benoist—a dead Yahweh or a vitalistic Wotan. In either case, to adapt a phrase from James Joyce, these are very posthumous gods.

For all their differences, Lévy and de Benoist have a lot in common. Each declares himself a moralist in philosophy, a nominalist in world view, and an antitotalitarian in politics. Both are skillful Parisian publicists (Lévy is an editor at Grasset, de Benoist at Copernic), and both have written much-acclaimed books (*Barbarism with a Human Face* won the 1977 *Prix d'Honneur de l'essai*, and *Vu de droite* won the 1978 *Grand Prix de l'essai* from the Académie française). Each has set flame to his recent past (for Lévy, Maoism, for de Benoist, the "Old Right") and risen like a Phoenix from the ashes to go on to condemn Marxism and modern liberalism, the Gulag and Coca-Cola, fascism of the left and right, the Inquisition, the Enlightenment, and the rule of the masses.

Yet as we might expect from these heralds of monotheism and polytheism, they have spent much energy excommunicating each other. There they were last year in the offices of *France-Soir* for a round-table discussion, glaring at each other uncivilly from their respective worlds, only a few days after Sartre and Aron had managed to shake hands over the issue of the Vietnamese boat-people. In the course of the exchange Lévy declared himself "shocked by the ideological and theoretical poverty" of de Benoist's writings, while de Benoist found Lévy's books "not worth a trifle." "I am filled with hatred for you," Lévy hissed. "I hate no one," de Benoist replied, for the sixteenth maxim of his code of aristocratic ethics (*Les idées* . . . , p. 52) enjoins: "Never hate, but despise often." It was the best show since Gore Vidal and William F. Buckley went after each other on television over a decade ago. The *nouveau philosophe* and the *nouveau droitier,* the prophet and the druid, seemed to deserve each other.

It is not easy to place Lévy and de Benoist in recent French philosophy, not least of all because it is stretching the word to call either of them a "philosopher." To be sure, Lévy studied under the Marxist Louis Althusser at the Ecole Normale Supérieure and claims to be a Lacanian. De Benoist, who studied law and letters at the Sorbonne, is an autodidact in the works of Nietzsche and Heidegger. The range of books that they cite is immense (but de Benoist, unlike Lévy, seems actually to read them), and the urgency with which they press their points would have you believe that the fate of the West hangs on the result of their debate.

Lévy, unlike de Benoist, is a child of the student revolution of 1968. After structuralism's Gang of Four—Lévi-Strauss, Foucault, Lacan, and Althusser—had "displaced" the human subject—the individual thinking consciousness—in favor of the linguistic code, and that subject's alleged history-making in favor of invariant structures, the revolt of May 1968 was a made-to-order structuralist's delight. More a cultural than a political crisis, more a synchronic liturgy than a diachronic historical event, it could be seen as reenacting the myths of the French tribe (1848, the 1870 Commune) around a transpersonal hero (the Eternal Child, *le révolté*) within neat classical unities of time and place (the Left Bank, May 3 through June 16). Although its political consequences were practically nil, this modern ritual did appear to prove what the structuralists had argued at some length: the supremacy of the code—in this case, the media—over the message to be codified. As cameramen freely crossed the barricades, ministering to both sides like priests in medieval wars, the essential point became clear: it is more important to *faire la une* ("make page one") than to win. The coverage of the event *is* the event.

The point, we may imagine, was not lost on the then twenty-year-old Bernard-Henri Lévy, who followed the action not in the streets but in his room, by television and radio, with a map of Paris across his lap. Without his skillful use of the press and television some seven years later, the so-called "New Philosophers" would never have been launched. In fact, Lévy, who is dramatically handsome and remarkably fluent, seems to have been made for television from the start (he acted in a TV film between writing his two books), even if it took him some years to get there.

After the debacle of May 1968, Lévy, then a Maoist, heeded André Malraux's call and went off to Bangladesh. There he awakened from his dogmatic slumber and discovered that there was no difference between "progressive" and "reactionary" corpses. After spending a week posing as a journalist in a group of lackadaisical guerrillas (they never fought), he took off to India where he got rolled by a junkie and, though the son of a millionaire, financed his way home by running booze between Bombay and Goa. Such enterprising skills, combined with his facility with words, served him well once he was back in Paris. One day he walked into Grasset publishing house, discussed some projects off the top of his head, and, *mirabile dictu,* got himself hired as an editor and, a few months later, was appointed the director of two new series of books. He corralled some manuscripts from old friends at the Ecole Normale, rushed them into print, and in 1976 took to the television screens to announce the birth of the "New Philosophers." A year later he crowned these efforts by publishing his own *Barbarism with a Human Face.* At that point he had more requests for newspaper interviews and TV appearances than he could conveniently handle, and he earned himself the title *pub-philosophe,* "publicity philosopher." Metaphysics, having long been dead and buried, was resurrected as a media hype.

The mood of the French press and public contributed to his success. The appearance of Alexander Solzhenitsyn's *Gulag Archipelago* in 1974 severely undermined residual sympathies for the Soviet Union, just as the later revelations about communist behavior in Cambodia shook liberal sympathies for Third World socialism. Moreover, the emergence of France's brand of Eurocommunism—permitting the alliance of communist and socialist parties in the Union of the Left—made many Frenchmen uneasy. The Common Program of the two parties, for example, called for government control over banking and credit. Since newspapers had been suffering the burden of rising costs since 1974, this was seen as an implicit threat to an independent and critical press. The collapse of the Union of the Left before and during the elections of March 1978 seemed to point up the hypocrisy of this uneasy marriage. As the Left's dominance of political discourse in France was increasingly shaken, the New Philosophers found a ready audience, not least among editors and television producers.

It is impossible to discuss the New Philosophers as if they represented a unified viewpoint on anything.[1] While they were all deeply affected by Alexander Solzhenitsyn's work, their only point in common may be that they have recently been issued by the same publisher. Some but not all were Maoists in 1968; one, Jean-Marie Benoist (not to be confused with Alain de Benoist), sat out the revolution as a diplomat in London, while another, Jean-Paul Dollé, fancies himself a Heideggerian. André Glucksmann, who publishes with Grassett but *not* in Lévy's series, refuses even to be grouped with them. Therefore, in discussing Lévy's two books (they have to be read together), I have no illusions that I am commenting on the other writers who are popularly associated with him.

Springtime, O. Henry once wrote, is the season when young men discover what young women have known all winter long. Lévy's bitter springtime, his discovery of the Gulag that other intellectuals, including Sartre, had known about for over twenty years, has engendered the purple prose, alternately threnodic and dithyrambic, that we find in *Barbarism with a Human Face* and *The Testament of God*. "If I were a poet," he writes, "I would sing of the horror of living and the new Gulags that tomorrow holds in store for us. If I were a musician, I would speak of the idiot laughter and impotent tears, the dreadful uproar made by the lost, camped in the ruins, awaiting their fate." This is pretty heavy stuff, but, as Husserl observed at the turn of the century, one is most vehement against those errors that one recently held oneself. "If I were an encyclopedist, I would dream of writing in a dictionary of the year 2000: 'Socialism, *n.,* cultural style, born in Paris in 1848, died in Paris in 1968.'" But Lévy is no easier on his young self: he confesses, with a straight face, "I will soon be thirty, and I have betrayed the dream of my youth at least a hundred times." Such earnestness is enough to make cynics weep, and it just might sustain some of them through the two hundred pages of narcissistic prose that one finds in his philosophical *Bildungsroman* called *Barbarism with a Human Face*.

Like the man in Paddy Chayevsky's *Network,* Lévy insists he is mad as hell and isn't going to take it any more. He has discovered, in a mood of "the darkest and most tragic pessimism," that the Marxism he once believed in is a lie: "No socialism without camps, *no classless society without its terrorist truth.*" Not that capitalism is any better. No, socialism is the face and capitalism the body of the same inevitable nihilism toward which the West is stumbling like a drunken Dimitri Karamazov. In fact, reality itself is radically evil, held in the clutches of an impersonal Power or Master or Prince or State (all in capitals and all equal to each other), as Plato and Schopenhauer, those "melancholy experts in absolute evil," knew. There is no Rousseauan nature that antedated the state and no revolutionary paradise to be found after the supposed "withering away" of the state. Nothing escapes the dread equation: World = Power = State = Barbarism. Misery will last as long as the social bond does, and that will go on forever. "Rebellion is unthinkable inside the real world."

But that leaves the "unreal world" and "the impossible thought of a world freed from Mastery." Thus, "the antibarbarian intellectual will be first of all a metaphysician, and when I say metaphysician I mean it in an angelic sense." In *Barbarism with a Human Face,* however, we come to the end without being told just what that might mean. Enter: *The Testament of God.* Its first principle is that politics must be restricted to make room for ethics and for an individual who can resist barbarism. Second principle: such an individual can *not* be found in classical Greek thought, where the individual is subsumed by the general and where the notion of conscience was unknown. It can only (third principle) be found in classical Judaism's "wager" on a Totally Other who is never incarnate in the world, in fact is now dead, although somehow goes on living, or partly living, in that "book of resistance" called the Bible.

The choice, then, is the same as it was for Tertullian in the third century: Athens or Jerusalem. Lévy's response is "Forget Athens." In place of its supposed humanism (which in fact is the root of totalitarianism insofar as it subsumes the individual under the general) Lévy proposes "seven new commandments." 1. The Law (Lévy's stand-in for God, but not to be confused with any specific *laws*) is outside time and more holy than history. 2. There is no eschatological future; rather, every moment is the right moment for manifesting the good. 3. The future is none of your business: act now. 4. Undertake no act that cannot be universalized for all men. 5. Truth, one's own truth, is extraneous to the political order. 6. Practice resistance, without a theory and without belonging to a revolutionary party. 7. In order to engage yourself you must first of all disengage yourself. If we ask Lévy what all this might entail for day-to-day politics, he comes down on the side of a "liberal-libertarian" state, which would govern best by governing least.

Little can be said about Lévy's position precisely because so little of it is ever argued. He makes his points by rhetorical tropes, wide-ranging historical references ("Consider the Middle Ages," he advises, or the span of history "from Epictetus to Malraux"), or by citations from books that he evidently hasn't read or has poorly digested (a reference to a work by Stalin in the Russian, which Lévy does not read, a reference to *all* of Clement of Alexandria's mammoth *Protrepticus,* which he has not studied, and so on).

He was taken to task in the pages of *Le Nouvel Observateur* in 1979 by Professor Pierre Vidal-Naquet of the Ecole des Hautes Etudes en Sciences Sociales for gross factual and historical errors: claiming that in *Genesis* Adam and Eve committed their Original Sin on the seventh day of creation (when God was resting), placing the action of Sophocles' *Antigone* in fifth-century Athens when in fact it deals with Thebes in the second millennium B.C. ("This," says Vidal-Naquet, "would be like using Racine's *Phèdre* as a document on Crete in the time of Louis XIV"), taking an 1818 text by Benjamin Constant as a commentary on an 1864 text by Fustel de Coulanges (Lévy in fact lifted both texts from a footnote in another work, but absolved himself of citing the source), and having Himmler stand trial at Nuremberg when in fact he had committed suicide on May 23, 1945. Lévy's sense of history is, to say the least, vague. When asked what he meant by saying that "the West was Christian even when the Scriptures were not read in the countryside"—and *analogously*—"The Greek world was Homeric even if, outside the Myceneaen palaces, the *Iliad* and the *Odyssey* were literally dead letters," Lévy confessed that he hadn't known that the Greek epic poems were written some centuries after the events they recount.

All this may be unfair. There is a long tradition of young scholars carrying out their education in public—Schelling enriched nineteenth-century philosophy by doing so. But it can be annoying when, instead of arguing his case, the young Dr. Lévy invites us, as he constantly does, to correct our intellectual errors by "reading" or "rereading" one or another major figure of Western thought, a task we might undertake if we thought Lévy had done as much. A rough count of his ABC of Reading includes: Lenin, Blum, Jaurès, the early Sorel, Plato's *Republic*, Marx's *Capital*, "the rules of the medieval convents," Rimbaud, Carl Schmidt, "the historians of the decline of the Hellenic world," *Mein Kampf*, Augustine's *Retractiones*, Nietzsche's *The Dawn*, St. Just, and Ernst Jünger. We are also encouraged to "go and see *The Night Porter, Sex O'clock, A Clockwork Orange*, or more recently *L'Ombre des anges*" in order to understand what harm has been wrought by Deleuze and Guattari's *L'Anti-Oedipe*. This makes one recall the quip attributed to Abraham Lincoln. "Better to remain silent and be thought a fool than to open your mouth and remove all doubt."

Alain de Benoist is a better writer, a clearer thinker, and a much more dangerous figure. He followed the events of May 1968 in the streets, but he saw them not as ushering in Year One of the New Order of Things but as a futile spectacle that announced "the end of the postwar period." While Lévy was off seeking adventure in Bangladesh, de Benoist stayed in Paris, tirelessly reviewing hundreds of books for the rightist publications *Valeurs actuelles* and *Le Spectacle du monde* (125 of these reviews were published in 1977 as *Vu de droite*) and seeing to the birth of the New Right.

De Benoist claims that the central issues of the traditional right, among them genetics, race, and inequality, have been discredited by their association with Nazism, and he tries to give them new life by grafting them on to such subsciences as sociobiology and ethnology. De Benoist is particularly attracted to sociobiology, which has recently gained an enthusiastic hearing in France. But he has a tendency to present the hypotheses of sociobiology as proven conclusions and then to extend these "conclusions" to far-ranging fields. For example, he writes, "all politics today implies a biopolitics." And he cites with enthusiasm the words of Professor Robert Mallet, the chancellor of the Universities of Paris, that some day "the genetic code will help inform the civil codes."

Although the French press and television woke up to the New Right only in March 1978, when Gilbert Comte ran a series of articles entitled *"Une nouvelle droite?"* in *Le Monde*, its origins reach back to March 1968, when the journal *Nouvelle Ecole* first appeared (de Benoist became its editor-in-chief in 1969) and to the founding, a few months later, of the study club called GRECE, an acronym for "Research and Study Group for European Civilization" (*Groupement de recherche et d'études pour la civilisation européenne*).[2]

Although de Benoist heralds these events as the beginning of a "new culture of the right," purged of the obscurantism, racism, individualism, and "father complex" of the reactionary right ("The Old Right is dead," he writes, "and deserves to be"), nonetheless the rosters of GRECE and *Nouvelle Ecole* read like a high-school reunion of old reactionaries and fascists. Jean Mabire, alleged collaborator in World War II and former editor of the extremist magazine *Europe Action* ("the magazine of Western man"), is now on the editorial committee of GRECE's newspaper *Eléments*. (De Benoist, who used to write for *Europe Action,* favorably quotes Mabire's paean to kamikaze pilots on page 227 of *Vu de droite.*) The *comité de patronage* of *Nouvelle Ecole* includes—besides such notables as Mircea Eliade, Konrad Lorenz, and Arthur Koestler—half of the editorial staff of the racist *Mankind Quarterly* of Edinburgh (R. Gayre, Robert Kuttner, and the late Henry E. Garrett) and at least two members of its Honorary Advisory Board (Bertil Lundman, a former contributor to the Nazi racist journal *Zeitschrift für Rassenkunde*—as well as H. J. Eysenck[3]). De Benoist himself is on the Advisory Board (and Arthur R. Jensen is an "Honorary Adviser") of the neo-fascist German magazine *Neue Anthropolgie,* whose editor, Jürgen Rieger, has condemned the "bastardizing" of races and has announced, in all seriousness, "The white giants are coming!" *Neue Anthropologie, Mankind Quarterly,* and *Nouvelle Ecole* all carry advertisements for one another.

GRECE and de Benoist have a strange penchant for the demi-monde of right extremism. On May 28, 1978, the *Washington Post* reported that representatives of *Nouvelle Ecole* participated in the eleventh annual conference of the allegedly anti-Semitic World Anti-Communist League in Washington, D.C. (its chairman, Roger Pearson, was formerly on the *comité de patronage* of *Nouvelle Ecole*) and met with William Pierce, a former spokesman of the American Nazi Party.[4] The Spring 1979 issue of *Nouvelle*

Ecole carried an article on pages 62–69 by one "Robert de Herte" (a collective pseudonym) on the inherited nature of musical talent. Footnote three on page 65 and footnotes six and eight at the end cite some thirteen works published in Nazi Germany on the topics of the "physical type" of great musicians and the relation between music and heredity. On May 29, 1973, GRECE sponsored a lecture on the theme of Europe by the self-declared fascist writer Maurice Bardèche, and de Benoist, in a chilling essay on *"Les corps d'élite"* in *Vu de droite,* approvingly cites Bardèche's remarks on "the exaltation of courage and energy" in Spartan education, followed by a rhapsodic description of the U.S. Marines by the rightist Francois d'Orcival.

The very powerful French publisher Robert Hersant—a former Pétainist who is currently the owner of one-fifth of France's newspapers—got into the picture when he bought up *Le Figaro* in 1975. He appointed Louis Pauwels—a well-known conservative editor who wrote an admiring book on Gurdjieff and was identified with the Gurdjieff movement—as the director of the spinoff weekly, *Le Figaro Magazine,* and Pauwels hired de Benoist to write a regular column on "the movement of ideas." Pauwels is also on the *comité de patronage* of *Nouvelle Ecole.*

Just what this all amounts to so far as de Benoist is concerned is still something of a mystery. Raymond Aron, himself Jewish, cautiously affirms that "Alain de Benoist defends himself from being [an anti-Semite], if not from having been one,"[5] but others have detected more than a whiff of racism in *Nouvelle Ecole's* fascination with the purity and strength of the Indo-European race. De Benoist, for example, finds it hard to conceal his enthusiasm for the French theorist of racial determination Arthur de Gobineau (1816–1882), whose *Essay on the Inequality of Human Races* asserted that different races have "very unequal destinies." Gobineau warned that Aryan society should resist mixing with the black or yellow races lest it lose its vitality and sink into corruption. De Benoist tries to salvage the *Essay,* which deeply influenced such French rightists as Charles Maurras, by calling it a work on the "diversity" rather than the inequality of races.

This much is sure: the one thing Alain de Benoist does not like is egalitarianinsm—not equality, which he takes to be an impossibility, but the *myth* of equality, the very idea that men should be equal. Not that he wants inequality *per se*. Rather, he wants diversity, "the right to difference," especially in racial matters, and with that a hierarchy, an elite, and a corresponding order, and, inevitably, then, *relative* inequality.

De Benoist does not believe, as Bossuet did not, that "some men are more men than other men," but he does agree with his colleague Pauwels that "equality is an injustice done to the capable." Nor is he a racist: all races, he says, are superior, and he is willing to go so far as to say that "all men of quality are brothers, regardless of race, country, or time." Although it is a fact, he says, that relative inequality comes with diversity, not all inequalities, especially of an economic sort, are just. De Benoist favors equality of chances (Nixon's Olympic metaphor of "an equal shot at the starting line"), and after that everyone is on his own.

Reading de Benoist's works, I had the clear impression that he did not arrive at his notion of inegalitarianism by induction from the data but that he began with it and then started collecting all the information that could support his conviction and attacking everything that might militate against it. The French have a pun: *Dis-moi que tu aimes, et je dirai qui tu es* (*hais*): "Tell me what you love, and I'll tell you who you are (whom you hate)." According to the sixteenth maxim of his code of ethics, de Benoist is not allowed to hate, only to despise (even though he delivers himself of the opinion that "one learns to *love* to the degree one learns to *hate*").

Nonetheless we can find out where his heart lies. De Benoist adores pagan polytheism because its many deities are made in man's image, consecrate his diversity, and guarantee his freedom. De Benoist depises monotheism because "its intrinsic totalitarian character" has engendered reductionism (where all knowledge can be led back to unity) and egalitarianism (which declares all men equal before God.) De Benoist loves the Indo-Europeans and especially the Celts for "their specific mental character," their physical characteristics, and perhaps (he cites Ernest Renan on the point) "the purity of their blood and the inviolability of their character." He despises Judaism (not Jews) for its intolerance and fanaticism, for consecrating a master-slave relationship before God, and for its "moral justification for killing the other." He likes biology because it affirms the diversity of species, and he despises Christianity, that "bolshevism of antiquity," which formed a counterculture of rootless slaves and Orientals who hated the very idea of fatherland, preached class warfare, and wrought "the progressive homogenization of the world" with their doctrine of universal love.

But fortunately for him the doctrine of equality has run through the three stages of its cycle—the mythic stage of Christianity, the philosophical stage of the Enlightenment, and the "scientific" one of Marxism—and the time is ripe to "raze the ground" and to start building the new myth of inegalitarianism. "We have something like a century in which to succeed," he writes, "which means that there isn't a moment to lose."

Preparing the ground for the new inegalitarianism entails educating an aristocratic elite of "supermen," not the muscular blond giants of Nazi fantasies, he says, but an elite of character. In a world that is intrinsically chaotic and meaningless and that gets its meaning only from the force of man's will, what are needed are "heroic subjects" who can create themselves and their own laws and who will remain faithful to norms they set for themselves. He cites examples from the motto of the Marines, *Semper fidelis,* as well as that of the SS, *Meine Ehre heisst Treue* ("My honor is called fidelity"). Such heroes will neither offer nor demand reasons, but will stick to their pledge and "keep silent." "Soldiers who, in order to fight, need to know why they are fighting are mediocre soldiers. And worse than them are soldiers who need to be convinced that their cause is good" (seventeenth maxim of the code of ethics).

In politics this translates into the "Organic State." Whereas today the state is no more than the sum of its inhabitants, de Benoist imagines a state that would be *more* than such a sum, and this "more" is called the *raison d'état* and is the basis for what he calls the "transcendence of the principle of authority." Precisely because America, dedicated as it is to "homogeneity" and "prosperous communism," does not understand these concepts, it "submitted the executive to the judiciary" and toppled President Nixon. And no wonder. "The very word 'fatherland' does not exist in the American vocabulary." No wonder, too, that America was defeated in

Vietnam. "The moving force in politics is not morality or phi-
lanthropy, but only energy. The essence of politics is energy. The
destiny of peoples is not shaped by 'interesting' cases or 'just' causes
but by the energy and force that are put at the service of these
causes—and at the service of others, to be sure." What might
motivate a nation to "serve others" is never specified.

It is not clear in de Benoist's case what is "new" about the "New
Right," any more than it is clear in Lévy's case what is "philosophical" about his "New Philosophy." De Benoist tinkers here and
there with the familiar model that calls for an elite based on the
superiority of white Europeans and is contemptuous of Christian tolerance and political democracy; but basically he serves up the same
old stuff. He styles himself a "raciophile," that is, one who wants
each race to preserve its own heritage and purity, as contrasted
with a "raciophobe," one who wants to blend races into a hodge-
podge. But behind this semantic subterfuge we still know who's
not coming to dinner. "We see some ideologues taking positions
on respect for all races—except one: ours (which by the way is
also theirs)," he writes. And citing Professor Raymond Ruyer of
the University of Nancy, de Benoist writes, "If one denounces, correctly, the ethnocide of primitives by Europeans, then Europeans
cannot be prohibited from protecting their own proper ethnicity
(*ethnies*)."

Such protection has a long history in France, and it should not
be surprising to find these sentiments coming to the surface at a
time when the rich and poor nations of the third world may seem
to impinge on Europe more ominously than ever before. What is
troubling is to find de Benoist getting a serious hearing and being
awarded a prize by the French Academy in the country of Mon-
taigne, who said, "Every man bears in himself the whole human
condition."

NOTES

1. The "New Philosophers" include Jean-Marie Benoist, *Marx est mort*
   (Paris: Gallimard, 1970); André Glucksmann, *Le discours de la
   guerre* (second, expanded edition, Paris: Grasset, 1979), *La cui-*

sinière et le mangeur d'hommes (Paris: Seuil, 1975), and *Les maîtres penseurs* (Paris: Grasset, 1977); Jean-Paul Dollé, *Voies d'access au plaisir* (Paris: Grasset, 1974), and other works; Guy Lardreau and Christian Jambet, *L'ange* (Paris: Grasset, 1976), and others. For a (not very helpful) critique see François Aubral and Xavier Delcourt, *Contre la nouvelle philosophie* (Paris: Gallimard, 1977).

2. Quite a separate phenomenon is the *Club de l'Horloge,* composed of some 120 young technocrats, most of them graduates of the Ecole Nationale d'Administration and the Polytechnique. Their spokesman, Yvon Blot, says, I believe correctly, "We have nothing to do with the New Right or with GRECE." However, M. Blot says, "Sociobiology is making spectacular progress. It cannot be ignored just because it is close to certain Nazi themes." On the *Club de l'Horloge* and the New Right, see *Le Matin* (Paris), July 25, 1979, pp. 15–17, July 26, 1979, pp. 10–11, and July 27, 1979, pp. 12–14.

3. On Eysenck see Peter Medawar, "Unnatural Science," *The New York Review of Books,* February 3, 1977, pp. 13–18. Eysenck is always careful to insist he opposes racial discrimination, but he also insists that "the contribution of genetic factors to variations in intelligence is something like 80 percent, compared with that of environment, which amounts to something like 20 percent." *Books and Bookmen,* September 1979, p. 48.

4. On the World Anti-Communist League, see Michael Billig, *Psychology, Racism and Fascism* (Nottingham: The Russell Press Ltd., 1979), pp. 25–26. Concerning Konrad Lorenz's early connections with Nazi ideas, see Bruce Chatwin's recent review, *The New York Review of Books,* December 6, 1979.

5. *L'Express,* July 21-27, 1979, p. 49.